Land-Use and Environmental Resources

Methods and Management

EDITORS

Dr. Anil K. Gupta is Head of Policy Planning Division, Head of Training Cell and Coordinator of Technical Advisory Committee, at India's National Institute of Disaster Management, New Delhi. Previously he was Director of Bundelkhand University Institute of Environment and Development Studies, and head of its Environmental Science and NRM Departments, as founder. He also worked with DMI Bhopal, NEERI (CSIR) Nagpur, NDMC Ltd. and CICON Bhopal. He is Fellow of Earth Scientists Society and known for pioneering contributions on disaster management policy planning, climate change adaptation, ecosystem approach to disaster reduction, environmental law and risk/damage assessment. He steered making of India's National Plan for Disaster Management, National Human Resource Plan, and Perspective Plan of NIDM. He was honoured with Young Scientist Award of Government of Madhya Pradesh (2006) and Excellence Award of Indian Society of Occupational & Environmental Health (2012). He has more than 125 publications including 20 books and 46 research/policy papers. He has steered various projects supported by NORAD, World Bank, Govt. Ministries, Planning Board, ICSSR, GIZ Germany, CDKN (UK), and collaborated with UNDP, UNEP (Geneva), UNU-EHS, ITC Netherlands, ISET (USA), inWEnt Germany, Ifanos Germany, etc. and represented in international delegations.

Sreeja S. Nair is presently associated with UNHABITAT Myanmar as Coordinator of the Myanmar Consortium for Capacity Development on Disaster Management (MCCDDM). She is Assistant Professor at National Institute of Disaster Management since June 2007. She is a disaster management professional having more than 15 years of experience. Her areas of research, documentation and training activities at NIDM include Geoinformatics applications in Disaster Management, Environmental Law, Disaster Data and Information Management, Ecosystem Approach to Disaster Risk Reduction and Chemical Disaster Management. She has published 20 papers in national and international journals, authored 8 training modules and authored/ edited 4 books. She has coordinated the Indo German Cooperation on Environmental Knowledge for Disaster Risk Management and was the co-principal investigator of ICSSR research project on drought vulnerability and mitigation analysis.

Prof. (Dr.) Mohammad Yunus is presently Vice Chancellor of the Mohammad Ali Jauhar University of Uttar Pradesh in India, since 2012, after his tenure as Professor (position offered in absentia) and Head of Environmental Sciences, and Dean of Faculties of Environmental Sciences, Information Science & Technology and Biosciences & Biotechnology (as Founder Professor), at Babasaheb Bhimrao Ambedkar (Central) University at Lucknow. He also served as Registrar and as Vice Chancellor of the University. He was elected Fellow of Lineus Society of London (FLS) and has the credit of scientific association with Prof. T. A. Mansfield (FRS) at Lancaster University, UK. He also worked at University of Sussex at Brighton, UK. Prior to joining as Professor, he was Scientist, Head and Area Coordinator of Environmental Science Division at CSIR's National Botanical Research Institute, Lucknow. He was also Convenor of Environmental Cell (financial arm for providing consultancy). Prof. Yunus has guided several doctoral and post-graduate research treatise in area of phytoremediation, land restoration, EIA, disaster management, etc., and published more than 140 research papers including 84 in foreign journals with high impact factor and high citation.

Land-Use and Environmental Resources

Methods and Management

Editors
Anil K. Gupta
Sreeja S. Nair
M. Yunus

2026

Associated Publishing Company®
A Division of
Astral International Pvt. Ltd.
New Delhi – 110 002

Cataloging in Publication Data--DK
Courtesy: D.K. Agencies (P) Ltd. <docinfo@dkagencies.com>

Land-use and environmental resources : methods and management / editors, Anil K. Gupta, Sreeja S. Nair, M. Yunus.
pages cm
Contributed articles.
Includes index.

ISBN 978-93-5124-089-1 (International Edition)

1. Land use--India. 2. Environmental management--India. I. Gupta, Anil K., 1970- editor. II. Nair, Sreeja S., 1976- editor. III. Yunus, M. (Mohammad), editor.

HD873.L36 2016 DDC 526.90954 23

Published by : **Associated Publishing Company®**
A Division of
Astral International Pvt. Ltd.
– ISO 9001:2015 Certified Company –
4736/23, Ansari Road, Darya Ganj
New Delhi-110 002
Ph. 011-43549197, 23278134
E-mail: info@astralint.com
Website: www.astralint.com

Laser Typesetting : **Classic Computer Services,** Delhi - 110 035

Acknowledgements

I recall of 1990s. As Member of the DOEN's Regional Environment Action Centred Task (REACT), along research studies, my work on water awareness for drought resilience in Central India made me to realize the importance of land resource and land-use management for human development and security. I, therefore, must express my gratitude to my teachers Prof. J. S. Rathore, Prof. Surendra Bhatnagar, Prof. G.P. Shrivastava, late Prof. V. S. Sai, Prof. Ugam Chauhan, and my M.Phil. & Ph.D. supervisor Prof. Rahasya Mani Mishra (former Vice Chancellor, APS University Rewa), for dilating my concern deep on environmental and land sustainability. I owe to Dr. J. S. Pandey, my guide at NEERI Nagpur (CSIR), and then coordinator of LUCC at IGBP, who exposed me to land-use issues, as part of carrying capacity based planning process.

Senior Editor of this volume, Prof. Mohammad Yunus, who mentored my learning in academic writing and publishing, has been a source constant encouragement and inputs to the making of this volume. I and thankful to my colleague and co-editor Sreeja S. Nair for all support and motivation. I am also thankful to my colleagues at NIDM, Prof. Santosh Kumar, Prof. Chandan Ghosh, Dr. Surya Parkash and Dr. K. J. Anandha Kumar for their encouragement and constant support. Motivation and support from Senior Professor at IIPA New Delhi and Vice-chairman of Sikkim Disaster Management Authority Dr. V. K. Sharma, Vice Chancellor of Devi Ahilya University Indore Prof. D. P. Singh, Director (Research) at IGNOU Prof. M. S. Nathawat, Project Management of NRM Unit of GIZ Germany India Dr. Rajeev Sharma, GGS Indraprastha University USEM Prof. J. K. Garg, former director of NIDM Dr. Satendra, and my former colleague/student Dr. Pallavee Tyagi, is gratefully acknowledged.

I am thankful to all the chapter contributors, and to Dr. Muralee Thummarkuddy, Chief of Disaster Risk Reduction at United Nations Environment Programme for writing the foreword, besides constant encouragements. Cooperation and support of editorial team of Astral International and Daya Publishing, and especially Mr. Anil

Mittal and Mr. Surya Mittal deserve special thanks. Last but not the least, family members of the Senior Editor (MY) and Co-editor (SSN), and of the undersigned (Dr. Alka Gupta, Ms. Vanya Badal Gupta and Mr. Vibhu Badal Gupta) are acknowledged for their cooperation without which this work would have not been possible.

-

Dr. Anil K. Gupta

Principal Editor

Foreword

Land is the foundation on which humanity has built civilisations and achieved progress. Land supports our forests and our agriculture. The quality of our urban and rural life is dependent on efficiency of our land use. It is also increasingly becoming clear that in appropriate land use planning can lead to disastrous consequences as evidenced by many recent disasters around the world.

The importance of land use planning for disaster risk reduction has been identified in a number of recent international documents. The IPCC Special Report on Extreme Events (2012) and the Sendai Framework on Disaster Risk Reduction (2015), both highlights the importance of land use planning for disaster risk reduction. The World Bank Report "Natural Hazards, Un-Natural Disasters", demonstrate land use planning as the most cost effective way of achieving disaster risk reduction.

While the importance of appropriate planning of land is both self evident and well documented, it is also a fact that adequate attention from the policy makers has not been directed in this area. Historically, when land was plenty, efficient use or even well planned use was not as critical as it is today. Secondly, as land is one of the fundamental possessions of individuals, families and community, any attempt to regulate land use will be seen as an encroachment of their freedom and way of life. Land Use planning is thus a major challenge, especially in democratic societies.

It is, therefore, important that the benefits of planned land use is studied, documented and used for policy advocacy at national, state and local levels. This book volume entitled "Land-use and Environmental Resources: Methods and Management" is a praiseworthy effort of bringing strategic issues along practical examples of useful tools and case studies on various aspects, from various countries. I congratulate the editors, who are well known to me as noted authors of text books and guidance documents, for undertaking this task. The publication is very timely in the light of global journey to Sustainable Development Goals. I am sure the book would be a significant reference for researchers, policy analysts, planning professionals, and post graduate students.

Dr. Muralee Thummarkuddy
Chief, Disaster Risk Reduction
United Nations Environment Programme (UNEP)
(Disaster Management and Post-Conflict Branch)
Geneva, Switzerland

Introduction

Land is the most limited resource as we have as yet only one earth known suitable for the human population which is ever growing and on a faster rate than the carrying capacity of the supporting systems. Land inhabits most life farms, holds water and water bodies, in-houses all the basic functions and precursors of climatic functions that take place in its near atmosphere. Changes in the land – landscape, land-use and quality, change the microclimatic settings and also modify regional and local influence on winds and precipitation.

Land is the house of all the economic developments – be it the housing, industry, power, agriculture, forestry, transport, or else, land is the basic requirement. Environmental resources are the systems and sub-systems of our environment (be it natural or human-made) that we look at for our services and supplies, also called as 'ecosystem services'. These are, for example, climate regulation, water, food and fibre, recreation, health and medicines, industrial raw materials, etc.

Management of environmental resources in an approach that helps maintain the renewability and sustainability calls for the effective management of land – landscape, land-use and land-resources. More recently the 'ecosystem approach to climate change adaptation' and 'disaster risk reduction' has added refinement to our understanding of 'environment-development' relations as 'human security and safety' are significant aspects of 'sustainability'. More so, the recent disasters of Uttarakhand (June 2013), Malingaon landslide (2013), Kashmir flood (2014), etc. have re-called an extension of the academic thought process to 'environment-development-disaster' inter-linkages.

In the modern times of growing urbanization, industrial agglomeration with urban peripheries and rural transitions in fast pace, defining natural and human environment separately would be inappropriate. However, land-use management holds the key and shall set the tone for management of environmental resources for human need and its generations. Case studies and experiences with examples from different parts of the world on different methods and management aspects of land and environmental resources are brought together in this edited volume with the aim of offering a holistic knowledge base covering all important aspects. The book comprises of three sections: Introduction and Overview, Methods & Case Studies and Management Strategies.

The Introduction and Overview section offers a broad understanding on land resources and land-use related current concerts of classification, land reforms and real estate, disaster risk reduction, and coastal specific issues. The section has 4 chapters contributed by practicing subject experts. First chapter of the book and of this section tries to relate the land and land-use functions with climate-change related issues and brings in the perspective of human security through land management. Land-cover modifications are more subtle changes that affect the character of the land cover without changing its overall classification. Sustainable land-use management provides a variety of cost-effective mechanisms for supporting both climate adaptation and mitigation. Ecosystem services represent the transformations of natural assets (soil, water, biota) into 'products' of value to humans. The context of land-use related researches offers emphasis on diverse range of studies on land-use which can be utilized for effective planning and management of ecosystem services with specific reference to climate-change adaptation.

Chapter 2 focused on the land capability classification as a tool for optimization and brings in the ground level research and extension experience of the authors. Six main uses of land have been defined; three related to ecological aspects, and three to human activities. The ecological uses, related to food security, compete with the human uses, related to urbanization, industrialization and recreation. Such competitions have had an adverse impact on this vital natural resource which is in fact the mother resource. The land-capability classification map is normally prepared by interpreting a standard soil-survey map. The land-suitability and its concerned capacity have been discussed in the chapter taking an example case study from Central India.

Sharing the accomplishments of a booming economy and healthy GDP growth, Indian real estate industry has been expanding at an exponential rate. Favorable demographics, rising purchasing power, availability of cheap finance, professionalism in real estate and reforms initiated by the government are the major drivers of this spectacular growth. Chapter 3 of the book presents a thorough analysis of land reform and environmental sustainability with special reference to real estate boom in India. Apart from economic development, employment generation and benefits of rural urban migration and number of other benefits can be counted of the real estate boom. But this reform encouraged environmental problem, social conflict and social imbalance in society. Land reform has widened the gap between rich and poor.

Land use planning in the coastal region through remote sensing based technique has been discussed in chapter 4 with a case study from Gujarat of India. Coastal areas are vulnerable in nature. At the same time they are also the growth engine of the region by attracting a variety of economic activities. Such a situation put lot of stress on the coastal areas for development. Gujarat example of coastal industrialization and land-change modifications in recent times is classic for study as it has multiple dimensions of benefits and losses to people and ecosystems, with warnings for the future. The change in land use pattern in the coastal region can provide an idea of utilization of land resources for development. The case study in the chapter also presents an analysis the quality of land in terms of salinity effects.

In the surroundings of Zaragoza (Spain), the quaternary alluvial terraces of the Ebro River deposited upon Tertiary gypsum formations. This covered karst area presents intense karstification processes and is an important source of sand and gravels deposits and groundwater. Negative interactions with the environment, as the destruction of infrastructure by land subsidence or groundwater contamination by industry, occurred in the last decades, due to the fast development of the city. Chapter 5 provides a case study of spatial decision support systems for sustainable land-use suitability analysis. The study the methodology was useful in classifying and identifying the land suitable for new industrial, agricultural and housing development.

Land cover and land use can be seen as the interface between natural conditions and human influence providing a sound framework for linking socio-economic development with some environmental impacts. Since independence, India has witnessed phenomenal educational development both in quantitative and qualitative terms. There has been a steady growth in school enrolment coupled with a decline in school dropout. Of all, land resource to house the educational buildings, playgrounds, hostels and residential quarters is the basic requirement. A study of land use for educational institutes, in a fast growing District of Uttar Pradesh in India, has been carried out and presented in chapter 6. Land under educational institutes, although occupies relatively small area (> 2% of the district) is important taking in consideration its potentiality of expansion in the crowded urban sprawling.

Chapter 7 provides an analysis of land use diversification in the Himalaya region, through a study of Shivaliks in India. The foothills of Shivaliks extending from Jammu and Kashmir to Uttarakhand states cover a total area of 3.10 mha. Rainfed areas face multifaceted problems of poor soil fertility, erratic rainfall, low soil moisture, poor economic condition of the farmers who cannot afford to invest adequately. This tends to diminishing livelihood support system. Land use diversification involves multidisciplinary approach including interaction among climate, soil, water, vegetation, livestock and socio-economic dimension in devising most remunerative, eco-friendly and environmentally sustainable land use. Since drought and floods are common evil in rainfed areas, appropriate techniques and technologies are to be adopted on watershed basis for effective soil and water conservation to have sustainable land diversification systems and improve the socio-economic condition of the poor farmers.

Recent flood disaster of Uttarakhand has re-iterated nature's warning of severe ecological backlash if our developmental planning fails to address environmental issues of the impact areas. Chapter 8 discusses the remote sensing and Geographical Information System application in environmental management planning of Vyasi Hydroelectric Project in Uttarakhand state of India. Environmental Impact Assessment (EIA) is a decision making tool to ensure consideration of environmental impacts of a developmental activity through incorporation of appropriate At present, various remote-sensing satellites are capable of providing extraordinary data on all aspects of the environment with microscopic resolution. These inestimable spatial data sets can be utilized in environmental decision making processes. Utilizing these abilities coupled with GIS, the tedious process of baseline data storage and subsequent analysis can be made efficient and less time consuming.

Development of Groundwater Prospects Zone Using Remote Sensing and Geographical Information System is the focus of chapter 9. The groundwater levels are decreasing over the years due to decline in annual rainfall, climatic changes, rapid urbanization, industrialisation, increasing runoff, agricultural activities and deforestation. The groundwater prospects evaluation for the parts of Hyderabad and Mahabubnagar districts of Andhra Pradesh has been attempted based on hydro-geomorphological mapping by using high resolution geo-coded merged satellite data (IRS-ID, PAN and LISS-III) in a GIS environment. In order to meet the future demands, less groundwater potential zones are identified and artificial recharge structures are recommended to improve the groundwater prospects.

Chapter 10 enumerates on groundwater resource management by hydraulic simulation of groundwater flow using Visual Modflow model. Ground water flow models for assessing groundwater potential have been a long-standing interest for water engineers. Numerical simulations based on a natural river channel indicate that explicitly modeling local obstructions/boundaries can significantly impact predicted flow parameters. A study on historical groundwater levels at confined and semi-confined aquifers in Vaigai river basin situated in Tamil Nadu showing a long-term fluctuation was undertaken. With the application of software like Visual MODFLOW, it has proved to be the convenient way to study such problems. In addition, it can generate large amount of data quickly and experimental modifications can be made with minimal efforts. These models are useful for better water resource management and understand the coastal hazards of the basin.

Water forms an essential commodity of life at domestic and ecosystem levels. It plays a complex and multifaceted role in both human activities and natural systems. Water Scenario in India in the context of the river-linking proposal of Indian Government during past decade is the focus of chapter 11. The river-linking has been put into water management discussion tables as a panacea of all hydro-meteorological troubles by some professional and political fractions without subjecting the matter to proper scientific and ecological scrutiny. The recent directive of the apex court in India also caused a situation of conflicts and dilemma regarding ecological perceptions of sustainable development and also the philosophy of 'win the nature'. Chapters 11 discusses the pros and cons of the water governance and context of riverlinking proposal from the angle of an environmental impact assessment.

Availability of water is highly uneven in both space and times but have competing uses. Land and land-use is crucial in water resource systems and its management. Chapter 12 discusses urban water supply system in context of water body management in an urban area. Reference has been taken to discuss the water use for drinking purpose. Planning and implementation of water resource projects involve a number of socio-economic aspects and issues such as environmental sustainability, appropriate resettlement and rehabilitation of project-affected people and livestock, public health concerns of water impoundment. Costs and benefit aspects of municipal water supply system has also been discussed with a perspective of a water balance.

Chapter 13 on 'agroforestry vis-a-vis environment' deals with alternative land-use promotion in ago-ecosystems particularly in rainfed and other drought prone areas of the country. Basically agro forestry is a land use system and the term was coined in 1970s. An ideal agro forestry system should possess the three attributes: productivity, sustainability and adaptability. Considerable work has been done in identifying the most suited system for the semi arid sub tropics where moisture evaporation is the double the average rainfall. Trees are a vital component of the survival economy of the rural poor, and this social reality has been utilized in promoting the agro-forestry models which are socially acceptable and ecologically suitable for the land.

– Editors

Contents

Methods and Case Studies

Management Strategies

Introduction and Overview

1

Land-Use for Climatic and Human Security: Policy Concerns

Anil K Gupta, Sreeja S. Nair & Mohammad Yunus

INTRODUCTION

Recent mega-disaster of Uttarakhand, 2013 flood that resulted in huge loss of life, human life in particular; infrastructure and environment, is one of the aloud warning of nature to human's ill-will. The inappropriate land-use not only in cities and towns but more importantantly on the mountains and their tributaries (rivers) – rivers catchments, slopes, basin and watersheds had caused the inferred natures havok. Land-use is probably one of the single most important aspect of anthropogenic environmental alteration contributing to disasters.

India has witnessed a growing trend in newspaper and television stories, almost daily, covering ecological devastations, land-use flaws, encroachments of natural systems and their consequences as aggravated floods, landslides, intense water scarcity or disease outbreak and several other challenges that our communities and public local administration face. However, the tragedy prevails. These efforts often fail to influence the conventional approach of planning and land-use decisions. The basis of this failure stems from the lack of political will and/or accountability in land administration.

'Land' is the most limited ecological factor on this earth which cannot be stretched to scale with growing population and their increasing aspirations. Implementation of economic and infrastructure development – be it at the state level, city level or befitting individual aspirations, land-use flaws continue to increased hazard-complex in which we now live and work. With vulnerable systems, resources and amenities, it marks a *big question* on 'sustainability and safety' of people and their environment.

A call to revive Delhi's water bodies captured in a daily newspaper (Hindu, March 29, 2013), as the attention drawn from a political party in opposition. The question is 'whether those outside' shall maintain the same concern while and after 'they are voted to power and to hold the authority'. Reportedly, this year 263 water bodies in South Delhi, South-West Delhi and North Delhi have dried up completely, aggravating water scarcity. There were a total of 1,022 water bodies under various government jurisdictions, of which only 219 could be saved. Seventy five water bodies are known to have become sewers and about 150 encroached. Legal or illegal constructions have taken over around another 800 water bodies. Challenges are aggravated due to heavy pollution of remaining water bodies and groundwater (Anon, 2013).

Another story (Dandekar, 2013) from the state of India crowned with the water success story of Ralegaon sidhi (*i.e.*, Maharashtra), where a town called Manmad is going without water for days. Running dry for over 50 days in a row, water conflicts have became a way of life there. Agitations and *dharnas* for water to survive, coupled with huge protests against 'water theft' by upstream farmers – a social crisis, which may turn to mega disasters. Police force was deployed along the canal, pumps were confiscated and cases were filed against over 200 farmers for lifting water. The tiny streets got clogged with cans and tanks when water tankers are about to come to fill these, amidst water fights, brawls and traffic jams. The water table falling rapidly leaving 300 bore wells sunk and with another 600 private bore wells. Lack of effective system and implementation from the state or local administration is the one aspect but the root of the present crisis lies in faulty land use systems in the region that disregarded water ecology in their practices. The grim situation in thousands of villages and towns, in a country which has excellent examples of Sukhumajri, Ralegaon sidhi, Tarun Bharat Sangh in Alwar, Jhabua water shed. and many more to its feathers, depicts a clear contrast of approaches and reiterate on 'drought of vision' in land use planning.

Land-use and Disasters

Climatic hazards such as floods, droughts, landslides, forest fires, pests and epidemics, have always been a primary matter of concern for human populations. According to the World Water Council, economic losses from weather and flood catastrophes have increased ten-fold over the past 50 yr, partially as a result of rapid climate changes. These rapid climate changes are seen in more intense rainy seasons; longer dry seasons, stronger storms, shifts in rainfall seasons, and rising sea levels. More disastrous floods and droughts have been the most visible manifestation of these changes (Eisenreich, 2005). Severe floods damage settlements, transport networks, and arable lands. Although devastating droughts are harmful primarily for agriculture and terrestrial ecosystems, they can also lead to local water supply and energy shortages. 'Avoidance' as a mechanism of 'Resistance' has been known in ecology since long. In disaster management – it is about avoiding 'high hazard prone' or 'critical areas' for permanent settlements/housing, establishing critical facilities, developing infrastructure or resources, etc.

Flood plan zoning has suggested an effective approach to avoid flood losses and at the same time help river maintain its originality and ecological health. Despite significant achievements in science and technology and success stories in environmental management in the 20th century, people still continue to suffer the consequences of climate hazards worldwide. However, evolving a 'land use approach' and 'land use models' that help reduce the occurrence and impacts of floods and droughts, or a landslide is still far from the understanding of our engineers or policy planners. This is a great lapse in their training and capacity building, and is a serious concern. Despite of enough literature and research existing on ecological aspects of environmental planning application for human security and such references at appropriate contexts in the national guidelines on disaster management (of National Disaster Management Authority, India), actual translation of these lessons to ground implementation is meager and uneven.

'Seismic micro-zonation' has been a buzzword of geo-technologists for earthquake risk reduction using land-use controls. However, how land-use alters the ground stability in the locality or site and thereby influences the liquefaction, has never been a subject of interrogation or planning concern – at least in India. So, most of us looking and applauding on the one side of coin, leaving other side unattended and unaccounted in the domain of disaster resistant housing. Standards/codes for earthquake resistant building existed for long (rarely implemented) but standards/codes for flood resistant buildings haven't yet framed. Risk of industrial mishaps / explosion leading to shock waves and heat waves and concerns of 'domino effect' of 'na-tech' (natural hazards triggering technological mishaps) disasters remains a gap area due to the hardcore construction engineers keeping away from environmental experts, and *vice versa*.

Land–The Most Limited Resource

All human activity happens on land because as yet we have not learnt how to build in ether and hence quite realistic that the land has to be acquired for each and every developmental activity. We, however very unfortunately have no framework of policy relating to land use and, therefore, both in matters of location as well acquisition, only ad-hoc-ism prevails (Buch, 2013). A retired civil servant and noted environmentalist, M. N. Buch in his article in Indian Express of 13th July 2012, quoted three very befitting example of 'lack of proper land use management policy', which caused government to face humiliation either by sabotaging the programmes, or even in a worst situation where a planned development programme had to be abandoned. Two of the examples are from West Bengal, *viz.*, Nandigram and Singur and the third one from Uttar Pradesh, the acquisition of land in Greater Noida.

The National Commission on Urbanization had reviewed the aspects of land use and management in detail. However, it noted data scarcity on how much land under actual cultivation had been acquired or privately converted to urban or non-agricultural use. The commission recommended for setting up of a Settlements Survey of India, duly assisted by the State Commissioner for Land Records, Survey and Settlements and a Directorate of Urban Land Records. It also recommended at national level a broad policy to be framed about what land is appropriate for what use and accordingly establish a hierarchy of land uses according to which land should be assigned. Detailed exercises could be done at the state , district and city levels to draw detailed development plans.

We need land for roads, railway tracks, airports, city expansion, industrial location, mining, afforestation, wetlands, tourism, etc. India's 60 per cent of land area is arable, approximately 30 per cent is under forests, grazing or other common use and only 10 per cent is uncultivable waste land. Locating non-agricultural activity on land best suited for agriculture must be restricted through the framework of a National Land Use Policy. In location planning, therefore, a national land use policy is of vital importance, and the strategy should be focusing on site selection for industries only but at the same time for land use for agriculture (including urban agriculture) green-belts, plantations/gardens and orchards, forestry, water bodies, waste disposal, temporary usage (recreational, sports, festivals, etc.) and for housing that is safe from disasters by virtue of location combined with design and construction. Thus, land use concerns and strategies need to be aimed and integrated to disaster risk reduction objectives.

Land-use Research

The land cover is defined by the attributes of the earth's land surface and immediate subsurface, including biota, soil, topography, surface and groundwater, and human structures. Data sets used in land-use/cover change research represent the land surface by a set of spatial units, each associated with attributes. These attributes are either a single land-cover category (*i.e.*, leading to a discrete representation of land cover) (Loveland *et al.*, 1999) or a set of values for continuous biophysical variables (*i.e.*, leading to a continuous representation of land cover) (DeFries *et al.*, 1995). Albeit concise and clarity of information provide the advantages of discrete representation of land cover, it unfortunately often led to an over emphasis of land-cover conversions and a neglect of land-cover modifications. Land-cover modifications are more subtle changes that affect the character of the land cover without changing its overall classification. Recently, there has been increased recognition of the importance of the processes of modification of land attributes (for example, agricultural intensification).

Impacts are not necessarily negative but are many times associated with reducing people's vulnerability and increasing well-being at the same time improving ecosystem services and integrity, *e.g.*, in certain cases for increasing food and fibre production. Numerous studies have been undertaken under the Land-Use and Land-Cover Change (LUCC) project of the International Geosphere-Biosphere Programme (IGBP) and International Human Dimensions Programme on Global Environmental Change (IHDP) (Turner *et al.*, 1995; Lambin *et al.*, 1999).

Global level studies estimated historical changes in permanent cropland during last 300 years by spatializing historical cropland inventory data based on a global land-cover classification derived by remote sensing, with a hind casting approach (Ramankutty and Foley, 1999), or based on historical population density data. The cropland area has increased globally from an estimated 300–400 million ha in 1700 to 1500–1800 million ha in 1990, a 4.5 - to 5 fold increase in three centuries and a 50 per cent net increase just in the twentieth century. The area under pasture—for which more uncertainties remain—increased from around 500 million ha in 1700 to around 3100 million ha in 1990 (Lambin *et al.*, 2003). These increases led to the clearing of forests and the transformation of natural grasslands, steppes, and savannas. Forest area decreased from 5000–6200 million ha in 1700 to 4300–5300 million ha in 1990. Steppes, savannas, and grasslands also experienced a rapid decline, from around 3200 million ha in 1700 to 1800–2700 million ha in 1990 (Ramankutty and Foley, 1999; Goldewijk, 2001). Europe, the Indo-Gangetic Plain, and eastern China experienced first the most rapid cropland expansion during the eighteenth century.

Land-use and Climate Risks

Climate processes are directly associated with the land use and land-cover, and therefore, has been a key research agenda during 1970s. It was recognized that land-cover change modifies surface albedo and thus surface-atmosphere energy exchanges with an impact on regional climate (Sagan *et al.*, 1979; Otterman, 1984). Role of the local evapotranspiration to the water cycle - that is precipitation recycling - as a function of land cover highlighted yet another considerable impact of land-use / cover change on climate, at a local to regional scale (Eltahir and Bras, 1996). As the research outcome, a better understanding of the land-use and land-cover's role in ecosystem services of climate regulation was generated. Concerns were primarily raised on land-use impacts on biotic diversity (Sala *et al.*, 2000), soil degradation (Trimble and Crosson, 2000), and supportive capacity of ecosystems (to human needs) (Vitousek, 1997). Land-use /cover changes affects the earth system functioning and poses local influence on global and regional climate processes including winds and thereby modifies resultant rainfall and temperatures. Land-use/cover changes is known to determine, to a significant extent, the vulnerability of places and people to climatic, economic, or socio-political perturbations including risks of associated natural hazards (Kasperson *et al.*, 1995).

The rivers of peninsular India are not perennial as they are not snow-fed. Health of these rivers is dependent on the forests of their catchment. A national land use policy is expected to zealously guard these forests and prohibit their conversion to other uses. A policy based on logic which any citizen could grasp, would ordinarily draw the public support for forest protection. We need land for grazing as a primary requirement for rural development and sustainability, but unfortunately many of our state governments have adopted policies which have virtually eliminated grazing areas.

Sustainable land-use management provides a variety of cost-effective mechanisms for supporting both climate adaptation and mitigation. Changes in land use such as deforestation and soil degradation - two devastating effects of unsustainable farming practices - emit large amounts of carbon into the atmosphere. Land-use change alone is responsible for approximately 20 per cent of global emissions of carbon dioxide each year. This has profound implications for the health and well-being of people and their ecosystems. Land-use sector has the potential role in global effort to address climate change - both by reabsorbing or preventing the release of carbon dioxide and by building robust ecosystems that support adaptation to the impacts of climate change. Sustainable land-management practices such as conservation of agriculture, intercropping and sustainable forestry can provide multiple benefits such as reducing erosion, building soil fertility and structure, improving water quality and buffering against drought and flood havocs.

Ecosystem Services

Ecosystem services represent the transformations of natural assets (soil, water, biota) into 'products' of value to humans. Ecosystem services have been defined as "the conditions and processes through which ecosystems, and the species that make them up, sustain and fulfill human life". The loss of ecosystem services will be a barrier to achieving the Millennium Development Goals (MDGs) to reduce hunger, disease and income poverty, especially in developing countries. Even if countries achieve the MDGs by 2015, the loss of ecosystem services will ultimately undo most of the benefits in the longer term. Although modern technology and knowledge can contribute to minimizing the human impact on ecosystems, their potential is unlikely to be deployed fully until ecosystem services cease to be perceived as unlimited and free and their valuation in economic terms is taken into planning and decisions.

Countries which provide most ecosystem services and preserve biodiversity for the well-being of global human population are rarely compensated for their efforts and contributions. For example, developing countries have committed themselves to the obligations of the Kyoto Protocol if they receive compensation for environment services. Markets for Ecological Services (MES) defined as mechanisms that create a market for ecosystem services in order to improve the efficiency in the way the service is used and these include the establishment of carbon sequestration offsets, tradable development rights, tradable quota systems, eco-labeling and environment-certification and bio-prospecting, among others.

One of the prime concerns of today is to understand the impacts of human activities on natural habitat affecting biological diversity and ecosystem services. Pressures can be physical (habitat alteration and fragmentation through changes in land use and cover), chemical (toxic contamination, acidification, oil spills, other pollution) or biological (alteration of population dynamics and species structure through the release of exotic species or the commercial use of wildlife resources). The main challenge is to maintain or restore the diversity and integrity of ecosystems, species and genetic material and to ensure a sustainable use of biodiversity. This implies strengthening the actual degree of protection of habitats and species, eliminating illegal exploitation and trade, integrating biodiversity concerns into economic and sectoral policies, and raising public awareness.

Concerns for Land Use Policy

There is a need for a national land use policy for land use planning and optimum utilisation of land resources. The linkage between land-use and water related policies and also within a broad framework of environment policy need to be made visible and effective, with drawing up further relevance to climate change and disaster risk reduction. Drought, floods, landslides, cyclones and epidemic disasters have direct implications of land use and land-water-forests-climate interfaces. Land-use policy must draw the guiding framework for its implementation through suitable tools and planning methods at national, state, district and local levels – broadly as a set of logical standards. A district level and local levels 'environmental action plan' must integrate 'land-use and land-cover' management within the 'integrated district plans' to be developed as guided by 73rd and 74th amendment in India's constitution (article 43) regarding participatory planning process. Concept of carrying capacity and cumulative impacts must be the guiding philosophy for land-use and developmental planning, especially in eco-sensitive and fragile areas (hills and coastal and river basin areas) and in certain cities and towns already under pressure of growth.

For example, in Madhya Pradesh under the old dispensation at least 15 per cent of the land in a village was reserved for *nistar* and, therefore, collectively formed part of the village commons. The present Government of M.P. in the footsteps of its predecessors is reducing 'village commons' by law, which is left virtually tnow not more than two per cent of the village area instead of 15 per cent reserved for *nistar*. A national land use policy need to reverse this and attempt to restore village commons so that at least the fodder and fuel requirements of a village are optimally met. Location of industrial activity on fertile land must be discouraged and uncultivable wasteland should be used for this purpose. Gujarat is the one state where government by way of policy has decided not to acquire fertile land for industrial purpose, but to encourage industry to locate in areas where land is totally not suited to cultivate.

Land system has to be viewed as a holistic resource and then assigned an appropriate use. A logical activity location policy and a framework within which agricultural growth would accelerate in tandem with urban development and industrialization, with the latter not moving forward at the cost of agriculture. All this is possible if we have a properly thought out national land use policy. The new land use policy needs to recognize the emergency land-use classes such as – urban agriculture, urban forests, which are greatly evident in Indian cities and towns, and contribute to the economy and livelihoods.

References

Anonymous, 2013. Call to revive Delhi's dying water bodies. *The Hindu*, Delhi, 29 March 2013, Page 4.

Buch, M. N. 2012.Chaos in Land Use Policy, The New Indian Express, 13th July, 2012. http://newindianexpress.com/opinion/article566386.ece

Dandekar, P. 2013. A drop in the bucket – Not a drop to drink. The Hindu, Delhi, 29 March 2013, Page 10.

DeFries, R.S., Field, C.B., Fung, I., Justice, C.O. and Los, S., 1995. Mapping the land surface for global atmosphere-biosphere models: toward continuous distributions of vegetation's functional properties. *J. Geophys. Res. Atmos.* **100(D10)**: 20867-82.

Eisenreich, S.J., 2005. *Climate Change and the European Water Dimension*. European Commission-Joint Research Centre, Ispra, Italy.

Eltahir, E.A.B. and Bras, R.L., 1996. Precipitation recycling. *Rev. Geophys.*, **34**: 367–78.

Goldewijk, K.K., 2001. Estimating global land use change over the past 300 years: the HYDE database. *Glob. Biogeochem. Cycles,* **15(2)**: 417–34.

Kasperson, J.X., Kasperson, R.E. and Turner, B.L., 1995. *Regions at Risk: Comparisons of Threatened Environments*. Tokyo: UN Univ. Press, 588 pp.

Lambin, E.F., Geist, H.J. and Lepers, E., 2003. Dynamics of land-use and land-cover change in Tropical regions. *Annu. Rev. Environ. Resour.*, **28**: 205–241.

Loveland, T.R., Zhu, Z., Ohlen, D.O., Brown, J.F., Reed, B.C. and Yang, L.M., 1999. An analysis of the IGBP global land-cover characterization process. *Photogramm. Eng. Remote Sens.*, 65(9): 1021–132.

Otterman, J., 1974. Baring high-albedo soils by overgrazing: a hypothesised desertification mechanism. *Science* 86: 531-33.

Ramankutty, N. and Foley, J.A. 1999. Estimating historical changes in global land cover: croplands from 1700 to 1992. *Glob. Biogeochem. Cycles,* 13(4): 997–1027.

Sagan, C., Toon, O.B. and Pollack, J.B., 1979. Anthropogenic albedo changes and the earth's climate. *Science,* **206**: 1363–1368.

Sala, O.E., Chapin, F.S., Armesto, J.J., Berlow, E. and Bloomfield, J., 2000. Biodiversity: Global biodiversity scenarios for the year 2100. *Science,* **287**: 1770-1774.

Trimble, S.W. and Crosson, P., 2000. Land use - US soil erosion rates: myth and reality. *Science,* **289**: 248–250.

Vitousek, P.M., Mooney, H.A., Lubchenco, J. and Melillo, J.M., 1997. Human domination of earth's ecosystems. *Science,* **277**: 494-499.

2

Land-capability and Land-Use Optimization

Hrittick Biswas, Brij Lal Lakaria & V.S. Katiyar

INTRODUCTION

Six main uses of land have been defined; three related to ecological aspects and the other three to human activities (Blum, 1993). The ecological uses, in particular related to food security, compete with the human uses of land, *viz.*, for urbanization, industrialization and recreation. This competition of land use had already had an adverse impact on this vital natural resource. It is on records that in the past half-century, about 2 billion, *ca.* 25 per cent of the 8.7 billion hectares (land for ecological aspects), namely agricultural land, permanent pastures, and forest and woodlands have been used for human activities.

In India alone about 105.5 M ha land has alredy been degraded (Ajai *et al.*, 2009). Accelerated water and wind erosion are the most serious cause of land degradation. This is supported by the fact that about 5,334 M tons of soil are washed away to ocean every year (Dhruvanarayana and Ram Babu, 1983), leading to a value of about 16.4 tons soil ha^{-1} $year^{-1}$. The need of the hour, therefore, is to protect the land from degradation and to restore/reclaim the degraded land using appropriate soil and water conservation measures.

Land Capability Classification

Soil and water conservation programmes, generally implemented on a pre-defined scale applying two distinct sets of operations, *viz.* (1) preparing a land capability classification map and (2) planning and executing measures to check erosion and improve productivity. The land-capability classification map is normally prepared by interpreting a standard soil-survey map. Land-capability classification (LCC) originally developed by the United States Soil Conservation Service in 1930s to interpret soil maps for farm planning, is a systematic arrangement of different kinds of lands according to those properties that determine the ability of the land to produce crops on a virtually permanent basis. Land capability classes are based on the intensity of hazards and the limitations of use. It helps in delineating a given area into units having similar constraints in management and productivity levels. It serves as a guide for the planners in deciding the effective conservation measures and predicting the production potential in respect of a given piece of land. The texture of the top soil, its effective depth, permeability of the top soil & subsoil, and associated land features like land slope, extent of erosion, the degree of wetness and susceptibility to flooding.

The grouping of soils into capability classes is done primarily on the basis of their capability to produce common cultivated crops and pasture plants without deterioration over a long period. The LCC includes eight classes of land designated by Roman numerals I through VIII. The first four classes are arable land-suitable for cropland-in which the limitations on their use and necessity of conservation measures and careful management increase from I to IV. The criteria for placing a given area in a particular class involve the landscape location, slope of the field, depth, texture, and reaction of the soil. The remaining four classes, V to VIII, are not recommended for cultivation, but may have uses for pasture, rangeland, woodland, grazing, wildlife, recreation, and aesthetic purposes. Within the broad classes are subclasses which signify special limitations such as (e) erosion, (w) excess wetness, (s) problems in the rooting zone, and (c) climatic limitations. Within the subclasses are the capability units (LCU) which give some prediction of expected agricultural yields and indicate treatment needs. The LCU unit is the most detailed category. Each LCU unit is defined by its assemblage of physical attributes: rock, soil, landform, slope, erosion, and climate (although other factors such as vegetative cover, land use and productivity indices are also taken into account). A LCU unit groups inventory map units together which respond similarly to the same management; which are adapted to the same kinds of crops, pasture or forest species; which have about the same potential yield; and which require the same soil conservation and other land management measures. Each unit is indicated by an Arabic number which follows the subclass limitation symbol (for example, VIe1). General types of conservation practices and uses recommended for land capability classes are listed in Tables 2.1 and 2.2.

A Case Study

a) Location

The study was conducted at the Research Farm of Central Soil and Water Conservation Research and Training Institute, Research Centre, Datia, spread over 191 ha. It is located at 26° 04′ N latitude and 78°28′ E longitude. The region receives an average rainfall of 860 mm with high degree of temporal and spatial variations. Eighty per cent of the annual rainfall is received during July to September. It is during this period that the soil is most prone to erosion hazards. Therefore, it was thought to

Table 2.1: Land suitable for cultivation.

Class	*Land Type*	*Sub-class (Dominant kind of Land)*	*Suitable for (Only the most intensive use mentioned)*	*Special Needs or Precaution*
I	Very good cultivable land (Light green on map)	Deep, nearly level productive valley land	Intensive cultivation to all climatically adapted crops	No special difficulty in farming; usual good farming practices to maintain soil fertility and conserve water
II	Good cultivable land (yellow on map)	II(e) Good soil on gentle slopes subject to water erosion or sandy soils	Cultivation with precaution	Protection from erosion; ues conservation irrigation methods
		II(w) Good soil slilghtly wet or subject to overflow	Cultivation with management of excess water and selection of crops adopted to wet conditions	Drainage improvement or flood protection
		II(s) Soil with minor soil problems such as clay or sandy texture, moderate depth or slightly alkali	Cultivation with selection of crops adapted to soil limitations	Treatments to offset soil limitations and to conserve irrigation water
III	Moderately good cultivable land (Red on map)	III(e) Good soil on moderate slopes subject to water erosion or sandy soil subjected to wind erosion	Cultivation with precaution against permanent land damage	Special attention to erosion control and conservation irrigation
		III(w) Good soil moderately wet or subject to overflow	Cultivation with careful management of excess water and selection of crops adopted to wet conditions	Intensive drainage improvement or protection from flooding
		III(s) Soil with moderate problems due to moderate depth, gravels, or alkali etc.	Cultivation with selection of crops adapted to soil limitations	Intensive treatments to offset soil limitations and to conserve irrigation water
IV	Fairly good land suited for occasional or limited cultivation (Blue on map)	IV(e) Moderately steep land subject to serious water erosion, or sandy soils subject to wind erosion	Occasional cultivation in rotation with hay or pasture, or orchards protected by permanent cover crops	Intensive erosion control when in cultivation

Contd...

Table 2.1–*Contd...*

Class	*Land Type*	*Sub-class (Dominant kind of Land)*	*Suitable for (Only the most intensive use mentioned)*	*Special Needs or Precaution*
		IV(w) Bottom land that is very wet or subject to severe overflow	Cultivation to special summer crops hazard of crop failure is always present	Intensive drainage, special attention to seeding and harvest dates to minimize crop failure on overflow land.
		IV(s) Fairly good land with limitations due to shallowness, gravel, stone, or strong alkali etc.	Occasional cultivation in rotation with hay or pasture	Very intensive treatments to offset soil limitations. Careful selection of crops
		IV© Good soil with just enough rainfall for crops in favourable years.	Cultivation during wet years, frequent crop failure; better in permanent vegetation	Conserve all rainfall; develop water for irrigation or convert to pasture or grazing use.

Table 2.2: Land not suited for cultivation

Class	*Land Type*	*Sub-class (Dominant kind of Land)*	*Suitable for (Only the most intensive use mentioned)*	*Special Needs or Precaution*
V	Very well suited for grazing, not arable (Dark green on map)	V(w) Good productive mountain meadows that are wet and have short growing season	Grazing and production of wild hay	Proper season of use and rate of stocking; protect form gulling.
VI	Well suited for grazing or forestry (orange on map)	VI(e) steep lands subject to erosion if cover is depleted	Grazing and/or forestry	Manage grazing and logging to maintain sufficient residue and litter on the soil for soil and moisture conservation; protect from fire
		VI(w) Flat land; occasionally with salin salts; permanently wet or subject to overflow	Grazing	Manage grazing to prevent soil puddling and to favour desirable forage plants
		VI(s) Flat to gently sloping shallow, stony, gravelly or alkali land	Grazing and/or forestry practices; fire protection	Good range and forestry management
		VI© Good or fairly good soil, not enough moisture for cultivation.	Grazing primarily; some forestry, could be cultivated if water available	Good range and forestry management practices; needs fire protection
VII	Fairly well suited for grazing or forestry (Brown on map)	VII(e) Very steep subject to erosion if cover is depleted	Grazing and/or forestry	Carefully manage grazing and logging to maintain enough plant litter for soil and moisture conservation; require fire protection
		VII(w) Flat, permanently wet or overflow land along streams, tidal marsh areas	Limited grazing	Range grazing management to favor desirable plants
		VII(s) Very shallow stony or strong alkali land	Grazing and/or forestry	Good range and forestry management, needs fire protection
		VII© Fairly good soil, not enough moisture for cultivation.	Grazing and/or forestry	Good range and forestry management practices; needs fire protection

Contd...

Table 2.2–*Contd...*

Class	*Land Type*	*Sub-class (Dominant kind of Land)*	*Suitable for (Only the most intensive use mentioned)*	*Special Needs or Precaution*
VIII	Suited only for wildlife, recreation and protection of water supplies (Purple on map)	VIII(e) Highly erodible – gullies bad lands and sand dunes.	Watershed and wild life	Maintain maximum cover for erosion control
		VIII(w) Tidal land, stream channels and swamps	Wild life, recreation and water spreading	Improve for wild life and recreation
		VIII(s) Barren mountain tops, little or no soil mantle	Recreation and watershed	Improve for wild life and recreation

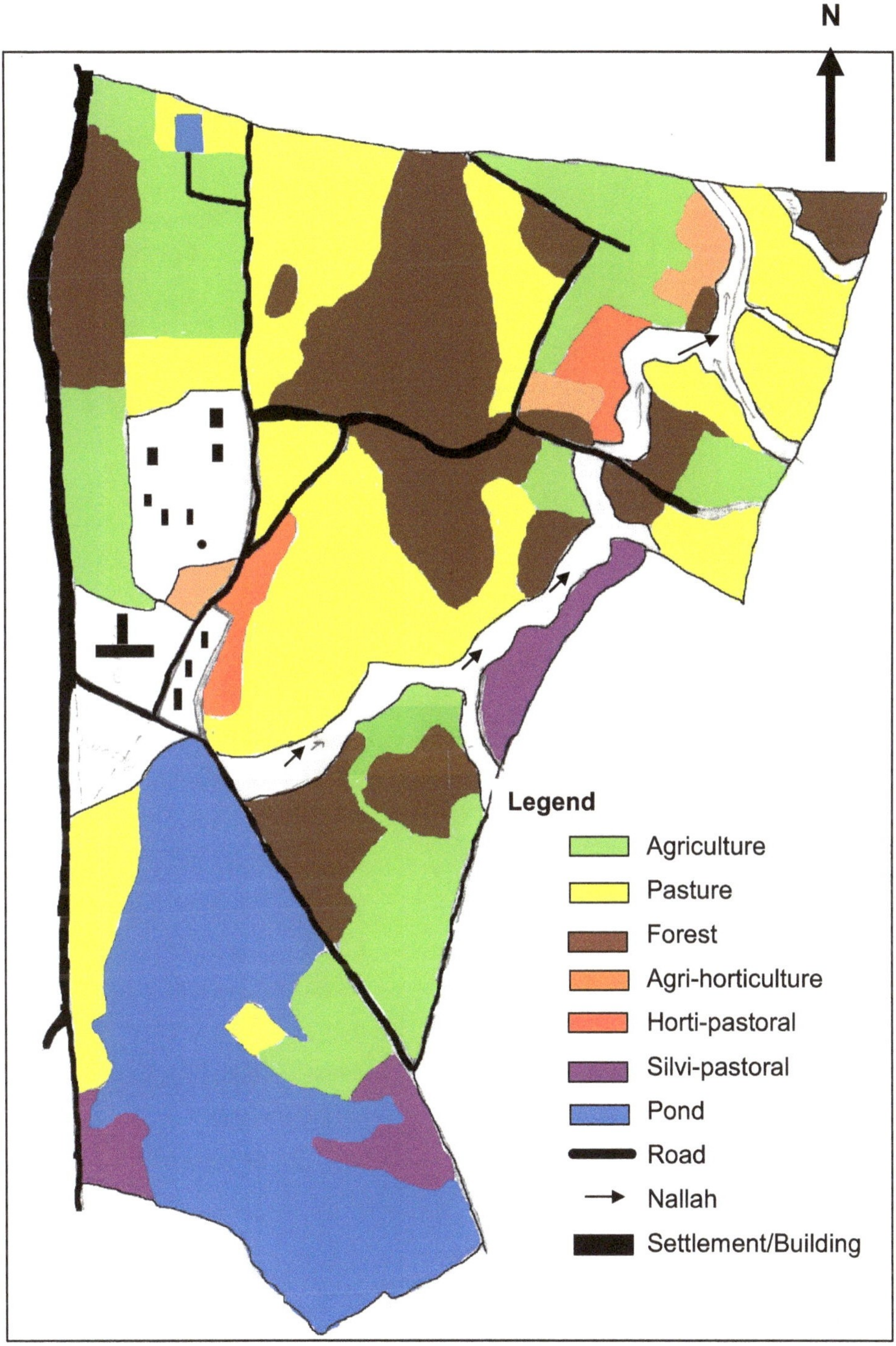

Figure 2.1: Present land-use map of the study area (191 ha).

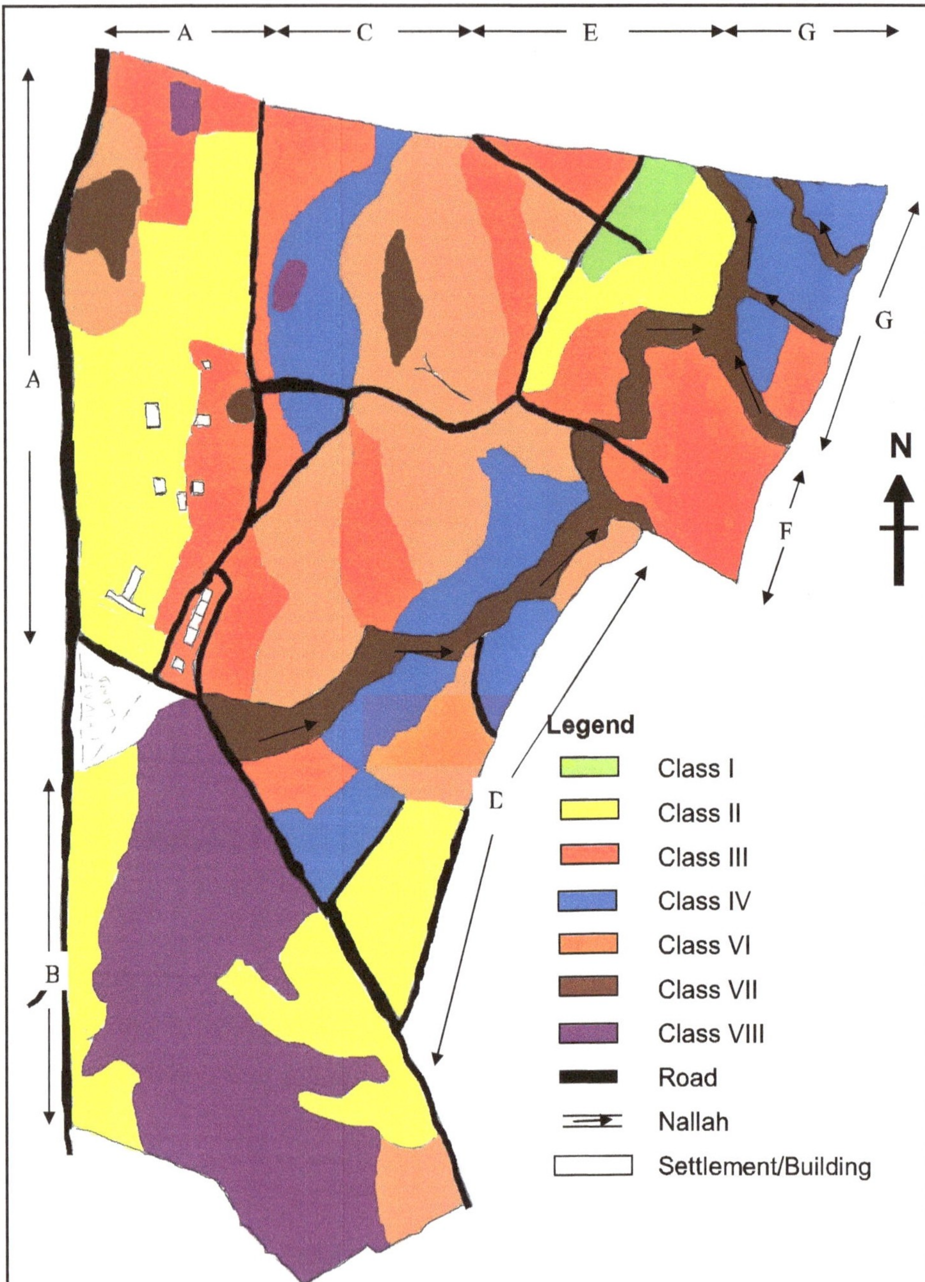

Figure 2.2: Land capability classification map of CSWCRTI, Research Farm, Datia (191 ha).

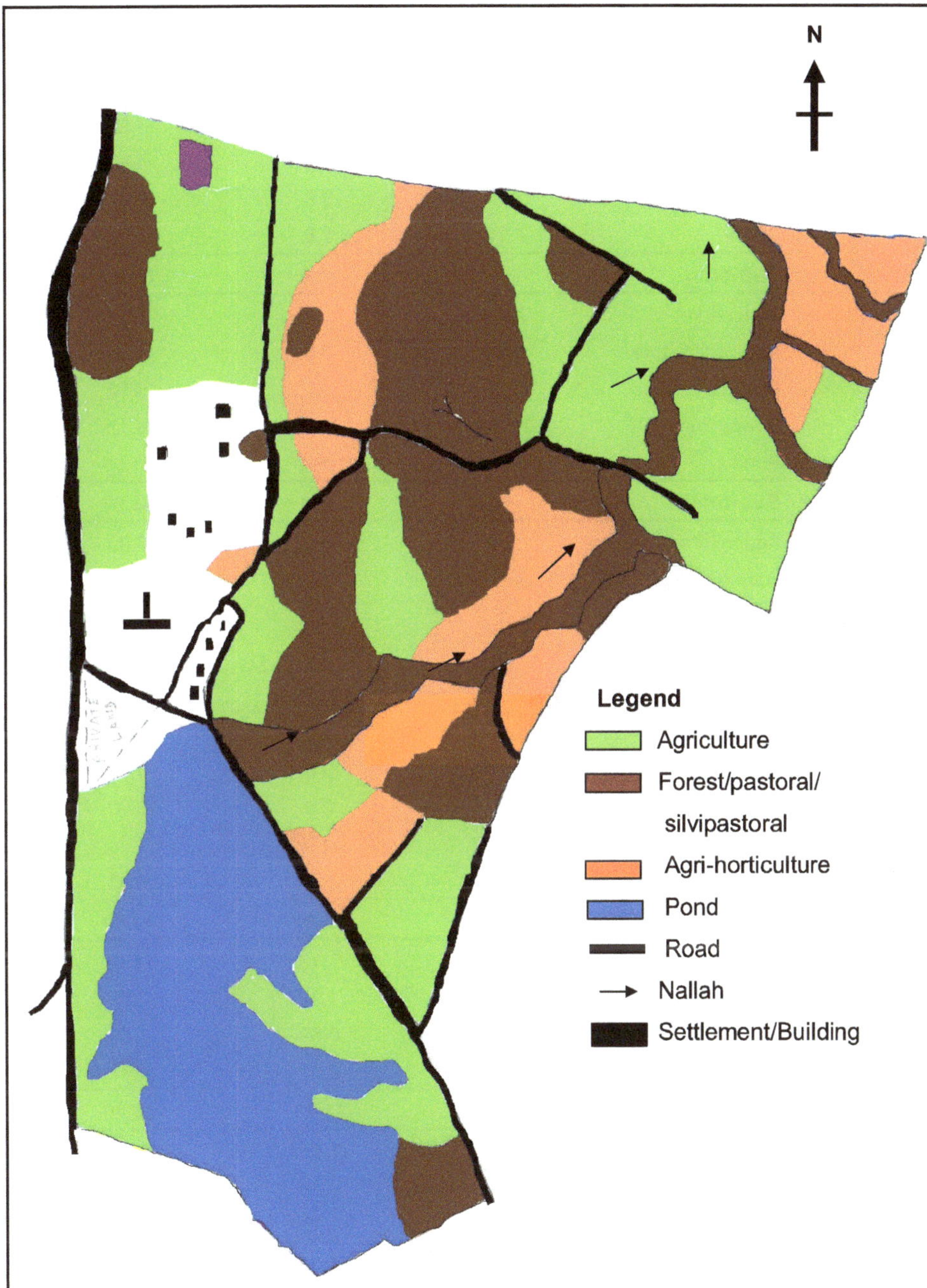

Figure 2.3: Proposed land-use of the farm.

Table 2.3: Present land-use of farm

Block	*Land-use*	*Block per cent*	*Farm per cent*	*Area of Farm*
A Block	Agriculture	29.7	5.6	10.75
	Agri-horticulture	2.3	0.4	0.84
	Pasture	20.0	3.8	7.22
	Forest	13.3	2.5	4.83
	Buildings	19.3	3.6	6.97
	Office	7.7	1.5	2.77
	Sub-total		**17.5**	**33.39**
B Block	Agriculture	7.0	1.7	3.19
	Agri-horticulture	0.0	0.0	0.00
	Pasture	11.3	2.7	5.12
	Silvipasture	11.3	2.7	5.12
	Pond	68.6	16.4	31.25
	Sub-total		**23.4**	**44.68**
C Block	Agriculture	0.0	0.0	0.00
	Horti-pastoral	9.9	2.2	4.20
	Pasture	50.6	11.3	21.50
	Forest	44.5	9.9	18.90
	Sub-total		**23.4**	**44.60**
D Block	Agriculture	42.6	4.2	7.98
	Agri-horticulture	0.0	0.0	0.00
	Pasture	0.0	0.0	0.00
	Forest	39.9	3.9	7.48
	Silvipasture	23.21	2.28	4.35
	Sub-total		**10.37**	**19.80**
E Block	Agriculture	63.28	7.12	13.61
	Agri-horticulture	11.72	1.32	2.52
	Horti-pasture	7.25	0.82	1.56
	Forest	21.91	2.47	4.71
	Sub-total		**11.73**	**22.40**
F Block	Agriculture	25.70	1.05	2.01
	Pasture	47.31	1.93	3.70
	Forest	26.88	1.10	2.10
	Sub-total		**4.09**	**7.80**
G Block	Pasture	90.48	3.86	7.37
	Forest	23.71	1.01	1.93
			4.87	**9.30**
Nallah			**4.71**	**9.00**

prepare a detailed land capability classification map of the Research Farm, the climate and soil of which are typical of Bundelkhand. It was envisaged that this study would be useful for planners and farmers of the region.

b) Present Land-use

Agriculture is being practiced on good to slightly degraded land of the Research Farm for more than twenty years now, Crops, *viz.*, black gram, green gram, soybean and sorghum are cultivated during *kharif* while wheat, gram and mustard during *rabi*. Pasture land use system is largely dominated by *Heteropogon contortus* (lumpa grass). Under silvipastoral land use system tree species, dhak (*Butea monosperma*), neem (*Azadirachta indica*), ber (*Zizyphus maurtiana*), prosopis (*Prosopis juliflora*) etc. do grow along with *Heteropogon contortus*. The hardy tree species, *Anogeissus pendula* (Kardhai) largely comprises the forest land use system in the rocky and gravelly terrains. The dominant land use systems in the Research Farm are shown in Figure 2.1 and Table 2.3. Among different land use system, agriculture (19.6 per cent), pasture (23.5 per cent) and forest (20.9 per cent) are dominant and cover about 122 (63.87 per cent) out of 191 ha. A significant area (16.4 per cent) is also covered under water resource, *i.e.* pond. Other land uses such as agri-horticulture, horti-pastoral and silvi-pastoral cover almost 15 per cent of the land area. About five per cent area of the Research Farm is occupied by buildings, roads, passages etc.

c) Survey Methodology

Extensive soil survey was carried out in the Research Farm with the help of a spiral soil auger, Abney's level and water bottle. The survey centered on four factors: (1) soil texture, (2) soil depth, (3) per cent and class of slope, and (4) degree of erosion. The classification of soils based on these factors is explained in the following sections.

Soil Texture

Soil texture was assessed through feel method in the field and were classified in one of the twelve textural classes as explained in Table 2.4 and the capability classes were assigned as shown in Table 2.5.

Table 2.4: Assessment of soil texture by feel method

Feel of Fingers	*Ballformation*	*Stickiness*	*Ribbon Information*	*Texture*	*Textural Classes*
Very smooth	Hard	Stains fingers	long ribbon	Fine	c, sic
Smooth	Moderately hard	Stains fingers	form but breaks	Moderately fine	sicl, cl, sc
Powery flour like	Firm	Stains fingers	Nil	Medium	l, sil, scl, si
Gritty	will shape but not form	Slightly	Nil	Moderately coarse	sl
Very Gritty	Nil	Nil	Nil	very coarse	s

c: Clay; s: Sand, si: Silt' l: Loam

Soil Depth

Soil depth was recorded by driving spiral auger to a maximum depth of 90 cm and the capability classes assigned to each depth range are shown in Table 2.6.

Table 2.5: Texture vs land capability class

Textural Class	*Capability Class*	*Textural Class*	*Capability Class*
s	IV	c	II/III
l	I	si	II
sc	II	sl	I
scl	I	cl	I
sicl	I	sil	I
sic	II	ls	I

Table 2.6: Soil depth vs land capability classes

Category	*Depth*	*Capability Class*
Very shallow (d_1)	< 7.5 cm	VI/VII
Shallow (d_2)	7.5 - 22.5 cm	IV
Medium (d_3)	22.5 - 45 cm	III
Deep (d_4)	45-90	II
Very deep (d_5)	> 90 cm	I

Land Slope

Land slope affects the rate of water intake by the soil, runoff and soil loss. The land slope was determined with the help of Abney's level and categorization into different classes was carried out. Values are shown in Table 2.7.

Table 2.7: Land slope vs land capability class

Description	*Symbol*	*Land Slope (per cent)*	*Capability Class*
Nearly level	A	<1	I
Very gentle slopy	B	1 – 3	II
Gently slopy	C	3 – 5	III
Moderate slopy	D	5 – 10	III
Slopy	E	10 – 15	IV
Steep	F	15 – 25	IV
Moderately steep	G	25 – 33	VI
Very steep	H	33 – 50	VII
Very very steep	I	>50	VII

Degree of Erosion

This qualitative feature in the field is based on experience and perception. However, as a guideline, Table 2.8 could be used for classifying soils into different units.

Table 2.8: Degree of erosion vs capability class

Degree of Erosion	*Symbol*	*Capability Class*
Slight/no erosion	e_1	I,II
Moderate erosion	e_2	III
Severe erosion	e_3	IV
Very severe	e_4	VI, VII

The symbols representing all the factors were expressed in the form of mapping units at all the locations surveyed in the following order:

$$\frac{\text{Soil texture - Soil depth}}{\text{Slope - Erosion}}$$

The distance between two points of survey may vary according to individuals perception, based on uniformity in texture, colour, slope or depth. However, once the map units are mentioned on different survey points, demarcation of areas is done based on the extent of a particular class.

c) Land Capability Classification and Proposed Land-use

Following the carried out survey, mapping units were demarcated on map (Figure 2.2). The area represented by each class is shown in Table 2.9. The land capability classification revealed highest proportion of land under classes III and II, covering about 46.6 per cent of the total Farm area. Area under Class I land is neglible. Thus cultivation in the Farm can be taken up with needful conservation measures. Maximum area that can be brought under cultivation in the Farm is 125 ha., *i.e.*, about

Table 2.9: Land capability classification of the research farm

Capability Class	*Area*	
	(per cent)	*ha*
I	1.45	2.76
II	22.89	43.71
III	23.71	45.29
IV	17.16	32.77
VI	10.71	20.46
VII	7.42	14.18
VIII	16.66	31.82
Total	100	191.00

64 per cent of Farm area (Class I to IV). Among agricultural crops, under class II and III land, gram and mustard may be grown. Near water sources like pond and tube well, where irrigation can be assured wheat may be cultivated successfully as highly remunerative crop.

Keeping in view the constraints, about 33 ha of land can be put under agri-horticulture for optimizing the production along with the adopting suitable conservation measures for erosion control. Fruit trees like aonla, lime, orange and guava are best suited for degraded land of this region.

Unfortunately around 66 ha of the Reaesch Farm area can not be utilized for agricultural/horticultural purpose. However this much area (66 ha) of the Research Farm is low lying and serves as a water reservoir (pond) which remains inundated during normal rainy season. This pond also receives all the runoff from nearby catchments outside the farm and serves as a regular/pertmanent source of irrigation for crops during rabi in particular. Only 34 ha land in the Farm is suitable for forest, pastoral and silvi-pastoral land use system comprising mostly of hardy species like kardhai (on hilly rocky terranin), neem, dhak, babool and prosopis spp. Along the water course and in the nallah bed, which covers about 9 ha, there is urgent need of carrying out soil and water conservation measures like check dams, gully control structures, spurs, vegetative barriers, afforestation etc. The proposed land use of the farm has been presented in Figure 2.3.

References

Ajai, A.S., Arya, P.S. Dhinwa, S.K., Pathan, K. & Ganesh Raj. 2009. Desertification/ land degradation status mapping of India. *Current Science,* **97:** 1478-1483.

Blum, W.E.H. 1993. Sustainable land use – a holistic approach. *Proceedings of the International Seminar on Managing Red and Lateritic Soils for Sustainable Agriculture – Extended Abstracts,* pp. 24-28, Bangalore, India.

Dhuruvanarayana,V.V & Ram Babu, N.P. 1983. Estimation of soil erosion in India. *Journal of Irrigation and Drainage Engineering,* **109:** 419-434.

3

Land Reforms and Environmental Sustainability: Reference to Real Estate Boom

Nisha Pandey & V. Vedak

INTRODUCTION

Sharing the accomplishments of a booming economy and healthy GDP growth, Indian real estate industry has been expanding at an exponential rate. Favorable demographics, rising purchasing power, availability of cheap finance, professionalism in real estate and reforms initiated by the government are the major drivers of this spectacular growth. The use of Modern geographic information system and information technology in modern land management brought dramatic change in the real estate economy in India.

The 1992 Rio Summit and subsequent literature and debate have focused on 'green' issues, such as biodiversity, climate change and marine pollution. Much less, however, has been written concerning the 'brown' agenda: those issues such as poor sanitation and water quality, air pollution and housing problems, which are particularly prevalent in the Third World cities. The Sustainable Urbanization in the Information Age address the global challenges posed by rapid urbanization and its impact on global warming and the natural environment - from poverty and inequality to natural and manmade disasters.

Objective of this study is to critically examining the progress of real estate economy in India and its impact on sustainable economic, social and environment development of society.

Growth Trend of Real Estate

The year 2006 started on a promising note when the Government of India opened the construction and development sector in 2006, and allowed 100 per cent foreign direct investment (FDI) under the 'automatic route' in order to spur investment in the vital infrastructure sector. The government has thrown open the lucrative parts of the Indian realty market to global investors for the first time. The relaxation of the FDI ceiling saw big names joining hands with the Delhi-based developments to announce India's largest FDI in the realty sector. The development of real estate in India focuses on two primary areas: retail and residential (www.omaxe.com).

The global real-estate consulting group 'Knight Frank' has ranked India 5th in the list of 30 emerging retail markets and predicted an impressive 20 per cent growth rate for the organized retail segment by 2010 (http://www.omaxe.com/boom_india.php).

According to industry estimates, the real estate industry in India has been growing at 33 per cent CAGR (compound annual growth rate) and could be a $50 billion industry in the next four years. The upturn straddles all the major sectors of the industry such as commercial, residential, retail, industrial, hospitality and healthcare (Real Estate in India http://www.iloveindia.com/real-estate/index.html).

The burgeoning outsourcing and IT/ITES industry have contributed to the demand for quality office space. With most of the blue-chip MNCs setting shops in the country, the estimated demand from IT/ITES sector alone is expected to be 150 mn sq.ft. of space across the major cities by 2010. Front-running companies like Wipro Spectramind and Daksh e-Services have been booming and hiring in the thousands in the last few months. The boom in commercial activity is being driven by IT and IT-enabled services. So, both commercial demand and residential demand, which is generated from the former, will continue to increase. Most of the development projects are to be undertaken through institutional finance rather than budgetary support (Amitabh Kundu, 2000).

Apart from economic development, employment generation and benefits of rural urban migration and number of other benefits can be counted of the real estate boom. But this reform encouraged environmental problem, social conflict and social imbalance in society. Land reform has widened the gap between rich and poor. Only rich can afford these multiplex and multicomplex structure, and poor can never be able to think about these place to live. It has created more slums with multi storied malls and female labours with BPOs and KPOs in big cities, poor sanitation and solid waste with high amenity banglows and flats in cities.

Few Dark Realities of Land Reform

Social Conflict

Indian Express Pune Newsline and Loksatta reported that thousands of farmers of village Maan near Pune battled with police, opposing acquisition of over 464 hectares of their livelihood farm land for the Rajiv Gandhi IT Park Phase IV. The land was being acquired at confiscatory rates. Two villagers were admitted to hospital with police bullet wounds, and many hurt in the lathi charge. Protests continue. The adjoining hills and 587 hectares of agricultural land from villages Hinjewadi, Marunji, Wasve Bhoirwadi and part of Maan had already been acquired by the state government for phases II and III (INDIA TOGETHER 7th April 2006).

Environmental Problem

Farmlands, public (government owned) wastelands, hills, riverbanks, ponds etc. in and around cities have been converted into habitation or infrastructure (roads, bridges, office premises etc.). This leads to:

- ✰ Reduced open spaces and interrupted air circulation causes 2-3°C higher temperature than outside the city, and electrical cooling implies coal emissions.
- ✰ Reduced soil surface due to higher built up areas has reduced water soaking ability of the city, causing weekly water logging or flooding of roads and home
- ✰ Gardens during monsoon, and stress of traffic jam reduces productivity.
- ✰ High air and noise pollution due to high travel distance between houses and
- ✰ Reduced absorption of emission or less oxygen supply due to loss of tree cover.
- ✰ Reduced biodiversity due to habitat loss, which is a cultural loss.

About 400 ton of organic waste gets generated in the Pune city, almost 150 g/head, daily, comprising of fruit and vegetable skins/roots, hotel leftover, sugarcane bagass, coir, flower waste etc. Inorganic waste is 60 per cent comprising mainly of paper, cloth and plastic, while 10 per cent of it is metal, glass etc. Each person generates about 0.45 kg solid garbage daily. Separate collection in/organic garbage efforts failed earlier due to civic inertia. Solid garbage is dumped into wastelands across the urban periphery, though some villages protested against it and may file lawsuit, as also in Bangalore. Organic manure making or herbal household biogas is not a lucrative business yet, but some 50,000 street children are involved in rag picking. Dharavi, a Bombay slum containing 6 per cent of the city population, generates Rs. 30,000 million, *i.e.*, 650 million US $ of goods annually-cloths, plastics etc. (Jamal, 2004).

Transport contributed 70 per cent of pollution here and other cities, except a bit less Kolkatta and Delhi due to manufacturing industries there. Delhi hosts electricity generating plants too. Supreme Court had warned 7 cities to reduce pollution to follow safety levels 2 years ago, with in vain. Air pollution exceeded safety level across 58 per cent of Pune city area, in winters (Anon, 2003).

Noise pollution is high- 90 decibel, and rarely 120-130 decibel, during the festivals due to firecrackers. Scrap units along the periphery - 500 of them-process city as well as imported waste - about 500 ton per anum or 1.5 ton per day the whole city. Such toxic waste causes lead or arsenic contamination, and explosions due to explosive material mixed (Jamal, 2004).

Thus we should link LR with environmental problem as supplementary forces of each other. Private players are encouraged to enter in these sectors. But in the long run there are two sides of entry of private players in this sector: at positive side entry of corporate bodies provides a platform for local citizens and extracts all the unutilized resources. Negative side of this step is that they tend to drain away local resources for their benefit and profit maximizing which may adversely affect the mobilization of internal resources

The ultimate aim of LR is human welfare and social security. This aims at the achieving higher economic growth with social justice. It is true in country like India; land reform can be meaning full and effective measure to eliminate all elements of exploitation of social injustice and to secure security and social justice through improved quality of life. Improved quality of life can be achieved by equal distribution of resources with equal property rights.

Land Reform in Theory

Land Reform (LR) usually refers to change in how land owner ship is distributed, recorded and administered. It occur in both the areas; rural and urban context. A common view is that land reform entails the redistribution of property right over land usually large land owner to peasants or land less rural laborers (Prasad, 2004).

Land reform was an important issue on the agrarian agenda in India after independence. Its main objective at that time was to eliminate feudal and semi-feudal relations in agriculture and strengthen the basis for broad-based and equity oriented agrarian growth through (co-operative) peasant capitalism (Srivastava, 2006).

In land reform we consider waste land in both rural and urban areas. Waste land is any piece of land which has been either damaged and has suffered degradation due to its over use, wrong use and uncared use or not being used. The need of reforms is there in both the areas; rural and urban, for socio-economic and environmental development.

Review and Historical Perspective

Economist distinguish two types of LR: a. Distributive or Reformist, b. Collective or Radical.

Distributive type is characterized by the redistribution of property right to beneficiaries. It usually takes place within a capitalist system. It is common to receive some payment as compensation for their expropriated assets, valued below the market price, depending on the extent of land reform. This type of reform was undertaken in Egypt, Kenya, Taiwan, Philippines, India and Pakistan (Ghose, 1983). Collective type is characterized by the establishment of producer cooperatives (collectives and cooperatives) and state framers. This is commonly undertaken by the socialist regimes. Usually all land lords are expropriated in such reforms and they hardly receive any compensation due to the revolutionary aims of new regime. This type of LR target land lords (owners of large landed estates), capitalist farmers who generally own smaller farms than those of landlords and who work the land much more efficiently using modern labor rather than tenant labors. Generally beneficiaries of such reforms are former tenants, rural wage laborers and small land holders. This type reform is far more radical than the redistributive type in terms of the greater number of land owners and the amount of land that is expropriated. This type of LR is being introduced in Cuba, Algeria, Ethiopia, China and North Korea (Sobhan, 1993). There are intermediary and variations of these two types of reforms termed as Hybrid type LR. Examples of LR in Tanzania in the late 1960 and in Chile during1964 -75 are better known.

As problem of poverty, landlessness and land conflict persisted in the country the agenda shifted in 1990 to a new version of land reforms as way of dealing with this problem with different perspective. This new concept of LR was referred as the' Market Led' LR. This is based on willing seller-willing buyer principle. The countries which have undertaken 'market led' LR are Brazile, Columbia and South Africa (Borras, 2003).

Contemporary Perspective

The structural reform and the associated development strategy formally launched in India in 1991 are expected to accelerate rural urban (RU) migration and boost the pace of urbanizations. The proponents of the strategy believe that linking India with the global economy will increase the inflow of foreign capital and facilitate indigenous investment, resulting in rapid development of industries. This is likely to give momentum to the process of urbanization since much of the industrial growth and consequent increase in employment will be within or around the existing urban centers. Even when industrial units are located in rural settlements, in a few years time, the latter acquire urban status (World Bank, 1995, 1998).

Land reform took place as a considered reaction in the global environment. The liberalization policy gives much accent on institutional reform and within this purview there is need to introduce major reform of agrarian institution. Newly marked terrains of economic activities across the world have called for examining the comparative advantages of nations in this sector.

When the reform occur in rural areas basically deal with agriculture and known as Agrarian Reform. Two schools of thought emerged to explain agrarian reform. First school of thought views the agrarian reform as a tool for bringing social equity through the fair distribution of rural income (mainly agricultural income) amongst the large number of peasants rather than amongst a few rich land lords. The second school of thought views the agrarian reform as a way of economic efficiency. It argues that agriculture production can be substantially increased by regulating of eliminating share cropper because the small farmers are more efficient in production per unit of land than big farmers (Lamba & Tomer 1986).

Objective and ways of Agrarian Land Reform (Rural Land Reform) are presented in Table 3.1.

Table 3.1: Objective and ways of land reform

	Economic	*Social*	*Political*
Objectives	Efficiency of production and Market growth	Distribution of wealth	Democratization
Method and Ways	a. Division of large land holdings.	a. Social equalization.	a. Democratization of society.
	b. Elimination of tenancy.	b. Mobilization of the cultural and political energies.	b. Providing satisfaction to the small farmers.
	c. Mobilization of production potential of agriculture.		c. Safeguards agrarian revolution.
	d. Diversification of production.		

Urban LR leads to reconfiguration of urban city with multi-complex and mall and can be termed as global city which are utilizing compressive advantage. In the era of global city and globalization comparative advantage has been change in to compressive advantage.

Urban Land Reform aimed for following objectives:

- ✰ Extensive use of land and labour
- ✰ Modernizations of city
- ✰ Strengthening the infrastructure of the country
- ✰ Opening of international standard offices, IT park.
- ✰ Generation of Govt. income, through tax(direct and indirect both), employment
- ✰ Improvement in basic services in religious, historical and tourist importance.

For a more systematic and coherent approach to reforms, developing a conceptual framework is required which traces the inherent characteristics of land and real estate assets to different types of market and institution failures.

The decade of the 1990s has also witnessed important changes in the thinking about cities and their role in economic growth and poverty reduction. There is a wider appreciation of the facts that (i) the size and nature of the urban problems that cities and towns face needs partnership between the public and private sectors. It is a gigantic task and any assumption that government alone can meet the infrastructure gap or provides housing to the vast numbers of the under-privileged groups is misplaced; (ii) the existing statutes need to be reviewed and changed as they do not permit expansion of the land and housing markets; (iii) the pricing system of urban services require major overhauling, as services when sold at below or above the cost, cannot be sustained (Baijal, 2007).

Current State of Land Reform

In the era of extensive structural and institutional reforms while most of the states are facing remarkable changes, some of them are suffering from great losses. UP, MP, Bihar have been lop sided due to failure of LR. Over the last five decades, the agricultural development in Bihar has been slow and tardy, pushing into the vicious cycle of backwardness and high incidence of rural poverty. At the same time although Punjab and Haryana have achieved remarkable growth in agriculture sector. The poor performance of agriculture and LR in backward states like Bihar, MP, and Rajasthan is not because of limited sources and scientific input but the structural and institutional deficiencies. This has worked as powerful barrier in the agricultural development of these states.

Real Estate as an Investment Avenue

There are immense opportunities for profit maximization and extraction of human resource by investing in real estate in the country like India. Positive demographic changes, cheaper and easier financing and regulatory changes have led to higher changes in investors mind sets. This reform is being seen as an attractive investment option, not only in metro cities, but even in second class cities is going to become attractive destination (TOI, 2006, Jan. 4).

Emaar being pioneer in regional real estate market have introduced the concept of gated communities, property purchase on freehold as well as innovative marketing and financing initiatives. Such partnerships are not merely about investment but also bring technology, expertise and scale. This joint venture will facilitate shopping, land escape garden, civic facilities, Technology Park, international offices which in turn generate employment and business. If properly directed such initiatives can bring infrastructure revolution (Galal & Rezzez, 2002).

New Real Estate Economy

The government has slashed cobwebs of legislation by allowing 100 per cent foreign direct investment in large projects. Stamp duties have been rationalized and reduced and a policy to set up special economic zones is boosting industrial growth.

Nascent chains such as Pantaloon Retail (India) Ltd. and Reliance Retail Ltd. have lined up multi-billion dollar expansion plans in which the largest component has been set aside for leasing or buying organized retail space. In addition, companies such as Nike Inc. and Adidas are deepening their presence through franchised operations beyond India's metros to soak up demand from millions in the growing middle class. Ambience currently has projects worth over Rs 850 crore (Rs 8.50 billion) in hand in Delhi and Gurgaon. All the newcomers are facing one problem: how do they make their presence felt when scores of new glass-and-concrete projects are already dotting the landscape?

India is coming as manna from heaven for a group of mid-sized construction companies. A few years ago they were the relative minnows of the construction business following in the footsteps of giants like DLF, the Hiranandanis and the Ansals. Today companies like Vatika, Senior Builders, and Ambience in Delhi or IDEB in Bangalore are taking on giant projects of their own and hoping to cash in on the real estate boom that's sweeping the country. Even in Mumbai the construction boom has pushed smaller companies into the limelight (December 13, 2003, Business standard).

The fact is that developers today especially the smaller ones have to throw in all kinds of features which they wouldn't have bothered with a few years ago. Evershine Builders, for instance, is providing a range of facilities from modular kitchens to piped gas and Internet connections. Some of its flats are even fully furnished. You just have to shift in with your family and clothes. Certainly, buyers today expect more. Closed-circuit television and earthquake proofing are also assured.

Is this a Real Estate Bubble?

The key to success in the real estate business, of course, depends on the price at which the land has been bought. That's where many mid-sized companies are at a disadvantage compared to the bigger groups. DLF, for instance, bought up large tracts of land in Gurgaon several decades ago. This put them in a particularly happy position and it means that many builders are also forced to buy from them.

Is this a building boom that will go on forever or is the music about to stop? Will Bangalore be the Silicon Valley of the east and the scene of unstoppable growth? And will Gurgaon be for India what Pudong, the giant suburb of Shanghai is for China?

Changing of land and its relentless move towards a market economy, will inevitable lead to further environmental problems also.

Structural Initiatives at the Central Level

Government has launched two development programmes; the Integrated Urban Development Programme and National Urban Development Mission. The Centre and state government's contribution could be in the form of a grant to the local body.

Government has launched two development programmes; the Integrated Urban Development Programme and National Urban Development Mission. The Centre and state government's contribution could be in the form of a grant to the local body.

Jawaharlal Nehru National Urban Renewal Mission (JNNURM), a gigantic scheme for urban development is being launched in recent past. The aim is to improve the basic services in over 60 cities with a population of more than one million, all State capitals and select towns of religious, historical and tourist importance. The scheme will involve a combined investment of Rs 1 lakh crore by the Centre, the State governments and urban local bodies over the next seven years. For this ambitious scheme to succeed, the State governments and the local urban bodies (LUBs) will have to reform the land market and the stamp duty structure. Many cities have archaic rent control laws and are tempted by the buoyancy of real-estate to use the stamp duty as a revenue-gathering exercise. This is going to be a major challenge in light of the fact that about 90 per cent of land in India is subject to legal disputes over ownership. Stamp duty ranges between 8 and 15 per cent of the property value, encouraging avoidance.

One more step has been taken by the Union Government to set up a National Urban Renewal Mission aimed at guiding and providing focused attention on urban infrastructure, the mission would help the urban centers in the country to realize their full potential and become engines of economic activity and growth. A national action plan has also been formulated to provide e-governance to all municipal bodies in a phased manner. The Centre has sanctioned Rs 5,000 crore in the financial year of 2005-06 and the first phase would be launched in August covering 35 cities with one million plus population. The Government had recommended municipal reforms to promote financial transparency and improve resource position of urban local bodies. This was one of the mandatory reforms to be carried out by State Governments and urban local bodies to access funds from the Centre under the mission.

The Government decided to link asset creation to the management of local bodies, as it was an important component for ensuring sustained service delivery. The draft note on the mid-term appraisal (MTA) of the 10th plan for the urban infrastructure sector has suggested major reforms in the property tax regime to ensure a major source of income for the urban local bodies and municipalities (Sing Manmohan, 2005).

Challenges in Implementation of Land Reforms

Followings are the other major challenges in implementing the land reform process in the urban areas:

- ☆ Improvement in the financial condition of local bodies in order to tackle the growing problems of urban poverty, shortage of housing and sanitation facilities.
- ☆ The cooperation and support of international agencies for achieving the eight Millennium Development Goals.
- ☆ The need for private participation in resolving the growing problems of cities.
- ☆ There is a need for a comprehensive scheme to provide assistance to local bodies for improving infrastructure, both in the small-to-large and metropolitan cities.

Recommended Policy Measures

1. A set of policy and regulatory reforms are recommended. It is estimated that removing land market barrier for FDI can contribute an additional 1 per cent to India's GDP growth rate (Mckinsey, 2001).
2. Metropolitan slump should be shifted by replacing them in homes for low and middle income group.
3. Urban and rural Municipal laws need further reexamination and regulation.
4. There is a great need to review and the modernization of latest development in infrastructure, transport, the urban pollution loads, etc.
5. All the tax related to land and property need to be reexamined.
6. The existing strategy for attracting FDI should be overhauled as the present urban land ceiling acts and rent control acts in states are serious constraint for the entire real estate sectors.

Conclusion

The actual economic and physical use of land and properties influences the land and its value. The land value is also influenced by the possible future use of land as determined by zoning and land use planning regulation. The land use planning and policies will, of course, determine and regulate the future land development. Land development should lead to an effective land use administration which would be the basis for sustainable approach to economic, social and environment development.

A global approach to land management depends on appropriate structures of governance. In this regard, the issue of decentralization could be significant key to achieve the general aim of sustainable development. Decentralization of land use planning and decision making requires synergy between local institutions and main governing bodies for managing these tasks.

In the context of sustainability, the conflict between immediate gains and needs on one hand, and the concern of future generation on the other is crucial. The principle of subsidiary is useful in considering the role of central and local government in land use management for sustainability. Environment sustainability requires action at all jurisdictional level.

There are global and transboundry environmental issues that need to be addressed at the local level. The regional and local level should be responsible for finding sustainable solutions to specific environmental problems due to new form of buildings. Policies can be implemented at different level and in different forms. Some areas such as pollution control at construction level and residential pollution are mainly implemented through vertical connections, by setting standards and regulations at national level to be administered at local level.

Rationalization in property taxes to boost economic activity in the housing and retail sectors is required, which would create immense employment opportunities. This is possible through overhaul of our land related laws, taxation and information systems. In currently existing system property tax rates are low and collection is inefficient.

The economic growth is desirable not only to change the living scenario and life style but also to reduce the environmental degradation caused by new life style and new forms of land reform. The present pattern of housing in mega cities is unsustainable with the currently available technology and hence this trend needs to be altered to limit environmental degradation. Price hike and labour exploitation in construction industry also needs to be controlled to be able to achieve a steady state sustainable development. If we act well and wisely and with charity, this is an opportunity we can seize to green the earth.

Sustainable real estate development and architecture are key considerations in environmental planning and management that seek to improve living conditions while preventing inequity and exclusion. The development of appropriate ideas, methods and clean energy technologies needs to be tested and applied.

Real estate development does not mean only quantitative transformation of land by new buildings but it is the result of both qualitative and quantitative transformation of whole society, a shift to a new ways of thinking and correspondingly, to new relation and new culture of society. This transformation qualifies only if people will improve their quality of life with shared vision of Vashudev Kutumbakam. Quality of life include happiness and satisfaction and can be termed as subjective well being of society but there is more in quality of life such as nutrition, a non hazardous environment and a long and healthy life.

The framework of analysis provided above describes the increasing importance of land reform to the national and global agenda from economic, ecological, and social perspectives. The direction of land politics and land reform in India will continue to be one of struggle and hope. It will be important to widen the scope of land reforms beyond the mere activity of redistribution of land or revisions of ceiling limits. In order to be effective, land reform must be seen as part of a wider agenda of systemic restructuring that undertakes simultaneous reforms in the sectors of energy,water and infrastructure and food security for social and economic and environmental sustainability.

The environmental impacts are best managed through coherent policy that set up properly legitimated rule systems and incentives. That policy applied to property rules as well as other economic incentives and sanctions can induce behavioural patterns that encourage sound and efficient use of resources. Environmental impact can be best managed when the basic needs of people are met-when addressing the problem of environmental security and sustainability.

References

Ahemed, Galal and Rezzez, Omer. 2002. *Real Estate Review* 2(1):00-00.

Anon, 2001. *Identifying the Barriers to Rapid Employment and Output Growth In India.* Mckinsey Global Institute, New York.

Anon, 2003. *Pune Environment Status Report for Municipal Corporation, Pune,* Department of Environment, University of Pune.

Baijal, Anil. 2007. Developing urban facilities. Ministry of Urban Development Press, Information Bureau Government of India, Bangalore.

Encyclopedia of International Development, 2005. Edited by Forsytn, T.I.M., Routtedge Pub., London.

Ghose, A. 1983. *Agrarian Reform in Contemporary Developing Countries.* St. Martin Press, London.

Jamal, N. 2004. Simmering city. *Down to Earth* **13(13):** 49-51.

Kundu, Amitabh, 2000. Management of Social Transformation (MOST), Discussion Paper No 48 Urban Developments, Infrastructure Financing and Emerging System of Governance in India.

Lamba, S.K. and Tomer, J.S. 1986. Agrarian reforms. In *Impact of Land Reform on Rural Developmen*. Agricole Publishing Academy, N.Delhi.

Prasad Jagdish, 2004. Land reform, an introductory view in the context of economic liberalization. *J. Social and Economic Studies* **16(1)**:00-00.

Sing Manmohan, 2005. Urban Mission. Bussinessline, Chenni, Dec, 6.

Sobhan, R. 1993. *Agrarian Reform and Social Transformation : Precondition for Development.* London.

Srivastava, S. 2006. Land, poverty, social justice and development. Draft paper, International Conference, Institute of Social Studies, The Hague, January 12-14.

http://economictimes.indiatimes.com

4

Land-Use Planning in the Coastal Region through Remote Sensing Based Technique: A Case Study from Gujarat

Shital Lodhia

INTRODUCTION

Coastal areas are vulnerable in nature. At the same time they are also the growth engine of the region by attracting a variety of economic activities. Such a situation put lot of stress on the coastal areas for development. With this regards, coastal region planning is very important as it provides strategies which strengthen synergies among the economic growth, its distribution and environmental conservation. Such strategies create linkages between various sectors. The role of land use planning is very significant in any regional planning. It is very important in terms of studying the existing trade-offs between various sectors and ensures that land is used efficiently for the benefit of the wider economy and population. At the same time it is also important to maintain the land resources in terms of their quality and productivity in a sustainable manner.

The geological and hydrological characteristics of the coastal region of Gujarat are highly distinct when compared to that of interior areas. Further, there is a large variation in geo-hydrological set up even within the coastal region. Although the coastal areas have developed substantially in terms of industrialization, agriculture has remained a major activity in some of the coastal talukas in Gujarat. Salinity due to the proximity of sea has remained one of the major obstacles for the development of agriculture in such talukas. Salinity in soil makes the region less productive agriculturally.

This paper tries to demarcate, assess and classify the land categories of Okha Taluka through remote sensing technique. It has further analyzed the quality of land in terms of salinity effects. The detailed analysis of the coastal land has been carried out by classifying the categories as per impact of salinity ingression. The remote sensing technique is an epitome to identify, demarcate and classify various types of land categories. Information on various categories play important role in land use planning for sustainable development. After assessing the quality of land in each category, recommendation for the suitable cropping pattern have also been recommended based on degree of salinity. The paper is divided into four sections. The first section assesses the existing land use and cropping pattern of the Gujarat Coastal region. The second section presents the methodology used for assessing the land categories of the selected coastal taluka. The third section discusses the results and also assesses the quality of land utilized in agricultural practices. The forth section suggest the cropping pattern as per the degree of salinity and recommend the strategies to improve the coastal livelihood.

Land-use Pattern in the Coastal Region of Gujarat

The change in land use pattern in the coastal region can provide an idea of utilization of land resources for development. To assess the change in land use pattern over a period of time, the area under different uses have been calculated. Later on, the change has been assessed by calculating the difference in area under particular use between two points of time. When the pattern of change is looked into the coastal region, twenty talukas have reported the change in their land use pattern over two decades. Among them seven have reported change from cultivated area to area not available for cultivation. This indicates that the land has been degraded and has become fallow mainly due to an increase in salinity. In order to assess the salinity affected cultivated areas, a detailed study of Okha mandal talukas has been carried out with the help of remote sensing technique. It is seen that the salinity is the main reason for the change in land use from agriculture to area not available for agricultural. The second type of change in land use pattern was observed in four talukas in which land use has changed from area not available for agriculture to forest or cultivated area. The main reason for such change is due to the plantation work carried out by many government and non government organizations. The third category of change is observed from cultivated to culturable waste land. This change is observed in five talukas. Development of small scale industries in these talukas has made such change in land use pattern. The fourth category of change is observed in three taluaks, where the area is transferred from forest to an area not available for agriculture.

This is because of the degradation of forest areas converting into land use for urbanization and industrialization. In short, the change in land use pattern in many coastal talukas is showing an unsustainable way of land utilization.

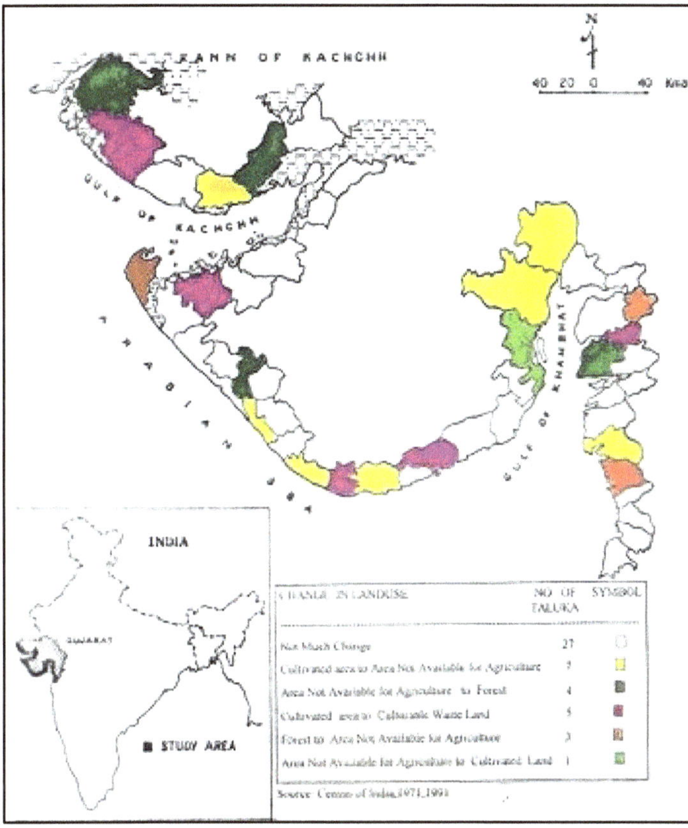

Figure 4.1: Change in land-use pattern in the coastal talukas of Gujarat.

Cropping Pattern in the Coastal Region of Gujarat

Agriculture is one of the major economic activities of the coastal region, as it provides food for population and raw materials for agro-based industries. The coastal region has 41,12,710 hectares agricultural land, which covers 22 per cent of the total agricultural land of Gujarat state. The region has very less percentage of an area under natural forest (4 per cent) compared to that of Gujarat (10 per cent). The reason for having less forest land may be due to both natural and edaphic reasons. The natural reason is mainly related to land degradation due to the marine ingression of salinity. The manmade reason includes mainly deforestation, which is again due to the great demand of land for urbanization and industrialization and also for fodder and fuel.

The region has 56 per cent net sown area, which is not very high when compared to that in Gujarat as a whole (which is 51 per cent). Similarly, the percentage of gross cropped area is less (107 per cent) in the coastal region when compared to that of the Gujarat state (115 per cent). This is mainly because of the lack of irrigation facility in the region mainly due to saline ground water. The following table shows the land use pattern and agricultural outputs of the coastal region as well as of Gujarat state.

Table 4.1: Land-use pattern and production of major crops

Sl.No.	*Parameters*	*Coastal Region*	*Total Gujarat*	*Percentage of Variables in Region to Total Gujarat*
1.	Agriculture Land (Hectares)	4112710.60	18822000	22 per cent
2.	Per cent of Forest area to total reported area	4 per cent	10 per cent	
3.	Per cent of Net shown area to total area	56 per cent	51 per cent	
4.	Per cent of Gross cropped area to Net shown area	107 per cent	115 per cent	
5.	Per cent of area under food grain to total area	26 per cent	38 per cent	
6.	Production of Cereals (in Tons)	1050827	3240054	33 per cent
7.	Production of Pulses (in Tons)	27912	537664	5 per cent
8.	Production of total food grains (in Tons)	1078739	3777718	29 per cent

Source: Census of India, Department of Agriculture 1991.

In spite of having very poor quality of soil and the dominance of cash crops, the region provides 29 per cent of total food grains production in the state. Among grains, the region mainly produces cereals. The region produces about 33 per cent of the total cereals produced in the state. While only 5 per cent of the total pulses in Gujarat is produced by the region. The low production of pulses makes the cultivated land more degradation because land gets back its nutritious elements like nitrogen and phosphorous through farming of pulses.

The region produces a large variety of crops. The pattern is very much similar to the cropping pattern of the state as a whole.

Table 4.2: Changes in area under different crops in the coastal region (1965, 1995)

Sl.No.	*Crops*	*Percentage Area Under Crops*	
		1965	*1995*
1	Cereals	37.00	20.19
2	Pulses	10.80	5.60
3	Fruits and Vegetables	0.72	2.51
4	Condiments and Spices	0.57	0.56
5	Other fruit crops	0.28	1.52
6	Oil Seeds	20.0	32.70
7	Fibre crops	16.0	16.17
8	Drugs and Narcotics	2.0	2.54
9	Fodder Crops	12.0	11.55
10	Total Sugar	0.63	6.66
	Total Area (in per cent)	100	100

Source: Agricultural Department, 1965, 1995.

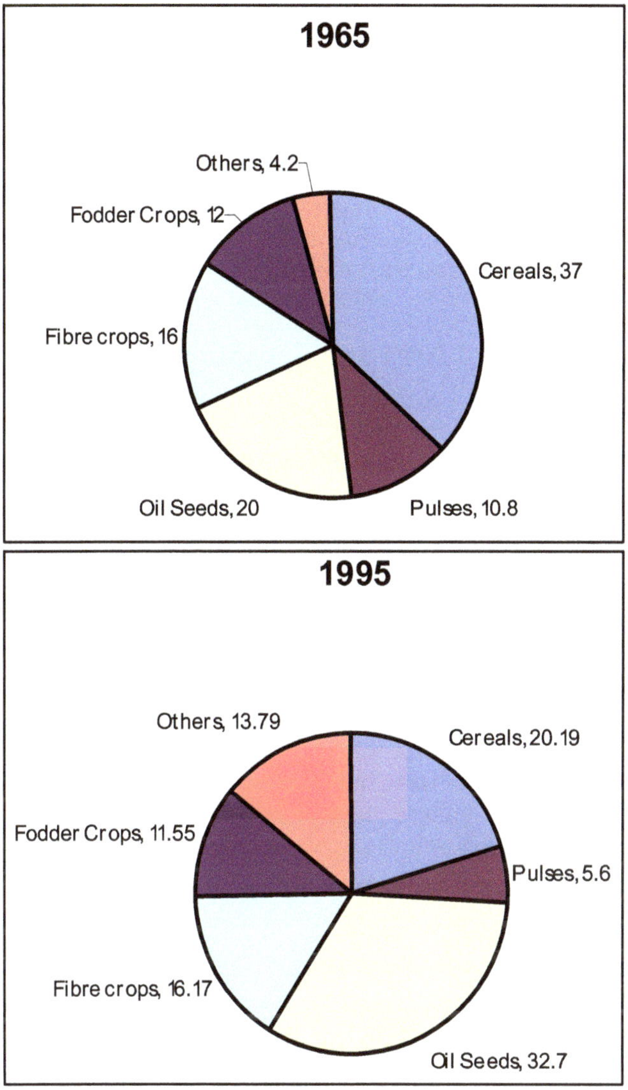

Figure 4.2: Cropping pattern (1965-1995)
(*Source*: Department of Agriculture, 1965, 1995).

The change in cropping pattern over thirty years of time period can be seen from the Table 4.2. The percentage of area under cereals and pulses have reduced over thirty years from 38.80 per cent to 25.79 per cent, while the percentages of area under oil seeds and sugar have increased over the same period of time (Figure 4.2). The percentage of area under oil seeds, which mainly includes ground-nuts for this region, has increased from 20 per cent in 1965 to 32.9 per cent in 1995. The percentage area under sugarcane has increased significantly from 0.63 per cent in 1965 to 6.66 per cent in 1995. Area under other crops has not changed much. Thus as far as the change in area under different crops is looked into, the cash crops have shown their dominance over food crops. The region has only 26 per cent cultivated area under the food grains. This less area under the food grains is a result of steady and progressive switch-over of farming from good grains to cash crops over these thirty years. Substantially high return value from the cash crops has made farmers to allot more land for cash crops cultivation. Among cereals, the region grows mainly paddy, wheat, bajra, juwar, maize, ragi, kodra and kutki. While in pulses, green grams, tur, black grams, masur, muth, val, and some other varieties are cultivated in a limited manner.

Salinity in the Coastal Area and Selection of Study Area

According to CSSRI (Center for Soil Salinity Research Institute), India has 7.0 million hectares of salt affected soil. Out of which 2.1 million hectares lie in the coastal tracts. Coastal region of Gujarat has 0.714 million hectares saline land which is 10 per cent of total saline areas of India. In short, the coastal region of Gujarat is highly affected by salinity, which has become one of the major hindrances in its development. The extent and magnitude of salinity varies within the region. This is reelected in terms of development. To study the problem with completely ground truth data is time consuming, expensive and inadvisable. Therefore, an attempt has been made over here to study the salt effected areas through technique of remote sensing in Okha Mandal taluka where salinity has deteriorated the cultivated land. Remote sensing technique is a scientific technique which provides detailed information of an object on the earth without coming into contact with it. It is one of the modern paradigms, which is widely accepted in almost every field because of its capabilities like synoptic view, repetitive coverage, availability of data handling system with improved sensors and economical viability.

Table 4.3: Location and characteristics of population of the study area

Taluka (District)	*Okhaq (Jamnagar)*
Latitude	22"00' N – 22"40' N
Longitude	68"45' E – 69"15' E
Population Density	201 people per sq.km.
Decadal Growth Rate of population (T)	22.75 per cent
Decadal Growth Rate of population (U)	27 per cent
Decadal Growth Rate of population (R)	15 per cent

Source: Census of India, 1991, 2001 and Survey Map of India.

The study area includes Okha Mandal taluka of the coastal Gujarat (Table 4.3). Although the taluka shows relatively better development, the region is large in area and shows greater variability in levels of socio-economic development, change in land use pattern and quality of ground water. The geographical and demographic characteristics of the taluka are shown in Table 4.3. It shows that although the growth rate of population is high the rural areas are not growing at par with urban areas. In fact, many villages have reported negative growth rates of population due to migration.

Methodology Used for Assessing Ingression of Salinity by Remote Sensing Techniques

The coastal area is continuously deteriorating because of the process of ingression of salinity. To find out the degree of deterioration in terms of salinity, remotely sensed data has been used. It was impractical to study the ingression of salinity in the total coastal area of the taluka within short period of time. The methodology used for this study is described here in detail.

IRS-1A LISS-II imageries have been enlarged to the scale of toposheets. (*i.e.*, 1:50,000) using the high magnification enlarger (HME). Toposheet and coastal landform maps prepared by RSAB, SAC have been used as base maps. The data of only one period will not satisfy the objective of classifying the areas according to the extent and magnitude of salinity. Therefore, both the data for pre and post monsoon periods have been used together. The extent and relative magnitude of salinity under cropped area could be classified based on the density of vegetation in both the periods.

For identification of different objects, imageries have been interpreted visually, taking the image elements like tone, texture, location, size and association into the consideration. Table 4.4 provides details of classification and their elements of image. Settlement in a lighter blue tone with rough texture, water bodies in black or blue tone with smooth texture, salt pans in a blue/white tone with square or rectangle shape, mudflats in a grey tone, sandy beaches in white tone with linear shape, salt encrust in a yellowish tone, barren land in a whitish tone and agricultural area with a red spot on greenish background have been delineated separately.

Agricultural areas can be divided into sub-categories on the basis of various parameters like topography, soil type, irrigation facility and degrees of salinity. Here agricultural areas have been divided into four categories with their relative magnitude of salinity based on the density of vegetation and encrust of salt. Density of vegetation was observed according to the different tone and textures. Four sub-categories of agricultural land could be classified.

Area Not Effected by Salinity

Area having dense vegetation in both the pre as well as post-monsoon period has been considered as a cropped area which is not affected by salinity at all. Dense vegetation shows a dominance of red dots on a greenish background.

Slightly Affected by Salinity

Area having dense vegetation in post-monsoon period but sparsely spaced vegetation with a lighter tone in a pre-monsoon period is considered as a cropped area, which is slightly affected by salinity.

Moderately Affected by Salinity

If during both the time periods sparsely spaced vegetation is observed with lighter tone, that area has been considered as a moderately affected cropped area by salinity.

Table 4.4: Elements of image interpretation

Sl.No.	*Category*	*Tone*	*Texture*	*Shape*	*Location*	*Association*
1.	**AGRICULTURE AREA**					
1.1	Cropped area	Red in both the pre and post monsoon	Rough	Irregular	Near settlement	-
1.2	Cropped area slightly affected by salinity	Dense Red in post monsoon sparsely spaced Red in pre-monsoon	Rough	Irregular	Near settlement	-
1.3	Cropped Area moderately affected by salinity	Sparsely spaced Red in both period	Rough	Irregular	–do–	–
1.4	Cropped area highly affected by salinity	Sparsely spaced red in post-monsoon and white in pre-monsoon	Rough	Irregular	–do–	–
2.	**FOREST**	Dark brown	Rough	Irregular		
3.	**WETLAND**					
3.1	Mudflat with Mangrove	Bright red	Smooth	Irregular	Near coast	–
3.2	Mudflat with other vegetation gray	Brownish blue/ brownish	Smooth	Irregular	Near coast	Between HWL & LWL
3.3	Mudflat without vegetation	Gray	Smooth	Irregular	Near coast	Along water line
3.4	Salt encrustment	Creamish dull white	Smooth	Patches/ meandering	Along HWL towards inland	–
4.	**WATER BODIES**	Black/blue	Smooth	Irregular	–	–
5.	**SETTLEMENT**	Light blue	Uneven	Radial or circular	–	Linked with roads or railway
6	**OTHERS**					
6.1	Barren land	Yellow	Smooth	Patches		
6.2	Stony waste	Grayish white	Rough	Patches	–	
6.3	Rocky coast	Dull white	–	Linear	Adjacent to coast	
6.4	Coastal Dunes	Red/White	Smooth	Linear	Towards land after the beach	
6.5	Salt pan	Dark/light blue or white	Smooth	Square or Rectangle	Near coast	
6.6	Sandy Beach	Bright white	Smooth	Linear	Adjacent to coast	

Highly Affected by Salinity

This area has sparsely spaced vegetation only in post-monsoon period. It is seen in dull white tone during in pre-monsoon period as no vegetation is observed there. This area is highly affected by salinity.

Mudflats with mangroves and other vegetation have been considered as moderately saline areas just because this vegetation prevents the marine ingression and makes the area less saline. Mudflats without vegetation, salt pan and salt encrust are considered as completely saline area. Barren land, stony waste, rocky coast, coastal dunes and sandy beaches are delineated separately as they are not saline. Table 4.5 presents symbols for each category derived from the above classification.

Table 4.5: Classification of the study area in different categories

Category	*Symbol*
AGRICULTURE AREA	1
Not affected by salinity	1.1 (Light Pink Shade)
Slightly affected by salinity	1.2 (Pink brown shade)
Moderately affected by salinity	1.3 (Light Brown shade)
Highly affected by salinity	1.4 (Dark Brown Shade)
FOREST	2
WETLAND	3
Mudflat with mangrove	3.1
Mudflat with other vegetation	3.2
Mudflat without vegetation	3.3
Salt Encrustment	3.4
WATERBODIES	4
SETTLEKENT	5
OTHERS	6
Barren land	6.1
Stony waste	6.2
Rocky coast	6.3
Coastal dunes	6.4
Salt pan	6.5
Sandy Beach	6.6
TALUKA BOUNDARY	—
HIGH WATER LINE (HWL)	—
LOW WATER LINE (LWL)	—

Source: Space Applications Centre, Ahmadabad.

Further, area under each category has been calculated for the taluka using dot grid method with the least count of 0.403 sq. km. Pie-diagram has been used to show the distribution of salinity affected agricultural area in this taluka.

Soil characteristics have been used to verify the accuracy of interpretation because these details are based on ground truth information.

Remote Sensing Data

As the study area is very large and scattered, collecting all the details by ground truth information is a time consuming work. The remotely sensed satellite data is an easy and quick method for this type of study. It provides synoptic and repetitive views of an area. Such type of view is accurate as well as convenient to collect in short duration of time. This study requires multi-date and historical data. Space application center (SAC of ISRO) has a facility of visual interpretation and historical data of the study area to assess such remotely sensed data. Therefore, the remotely sensed data have been used for assessing salinity affected areas in the selected taluka of the coastal region. Pre-monsoon and post- monsoon information of the study area is a minimum requirement for this study.

Following data have been used in this study for different purposes:

a. Satellite Data
b. Topographic maps
c. Land Form maps
d. Soil maps

IRS-1B, LISS-II remote sensing imageries of Pre-monsoon and post monsoon period were used for delineation of salt affected areas and other landforms.

Table 4.6: Satellite imageries and topographical maps used for Okha Mandal taluka

Satellite Imageries				
Path	*Row*	*Quadrant*	*Acquisition Data*	
			Pre-monsoon Dates	*Post-monsoon Dates*
34	52	A2	13 May	05-October
35	52	B2	31- May	28-October

Topographical Map

Topographical Map	*No*
41	F/4 & B/16
41	F/3
41	B/15

Source: Space Applications Centre, Ahmadabad.

Topographical maps of Survey of India have been used as base map of the taluka. Table 4.6 gives detailed information of path, row and acquisition date of imageries as well as the numbers of topographical maps used for this study. Coastal land-form maps prepared at RSAG (Remote Sensing Applications Group), SAC have

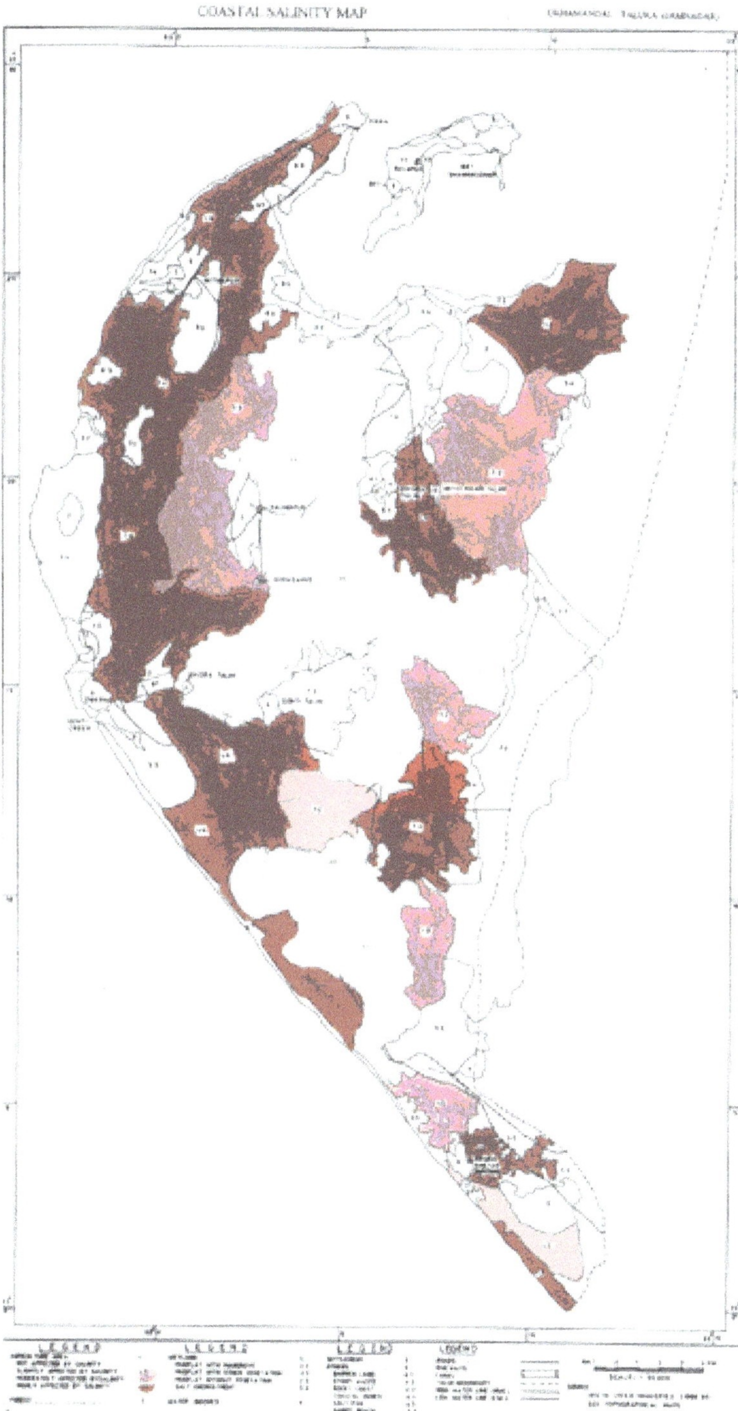

Figure 4.3: Map showing salinity classification in Okha Mandal Taluka.

also been used as ground truth data. Data about soil characteristics have been collected from Agricultural Department and various other reports. These ground truth data have been used to check the accuracy of remotely sensed data.

Analysis of Soil Salinity

The salinity within the taluka varies along with the distance from the sea. The selected taluka shows different pattern of salinity in terms of extent and relative magnitude.

Salinity in Okha is observed to be in circular pattern. This is because the taluka is surrounded by sea from three sides. Salt encrust, barren land, salt plan and highly saline cropped areas are observed very near to coast making an inner surrounding belt of 10 to 15 km (Figure 4.3). Salt encrust is observed in an area above Dwarka and in a narrow strip of 1 to 2 km. in the eastern part of the taluka. Highly saline cropped area is also observed near water bodies which are connected to sea through a water channel. They are present in the area between Mithapur and Dwarka, in surrounding areas of the Dwarka up to Gomti talav and surrounding areas of Bhimganj and Mithakhali lakes in the upper eastern part of the taluka; and also in the surrounding area of Dhora lake near Okha desert. Moderately saline areas are present in the eastern part of the taluka as well as in the inner belt adjacent to highly saline belt in the western part of the taluka. Slightly saline areas are located in the surrounding area of Dhinki and also in the lower middle part of the taluka. The central inner parts as well as areas around Navi Dhrewad are the only agricultural areas, which are not affected by salinity.

Salt pans are observed to be present near Mithapur, Armada, Padli below Okha in the north-west, in the areas located in upper part of Rajpara in north east and near Okha desert located in southern part of the taluka. Mudflats are observed near Gomti creek in west and in an area of southeast part which is known as Okha Rann. Reserved forests are delineated near Mithapur, Rajpara, Dwarka in northern part and near Kuranga in southern part of the taluka. Plantation is observed in a narrow belt between Kalyanpur and Dhrasnvel. Overall pattern of salinity in Okha looks like a concentric zone model. Degree of salinity decreases from the surrounding circular belt to the central part of the taluka. This very clearly indicates the impact of marine ingression on the quality of land.

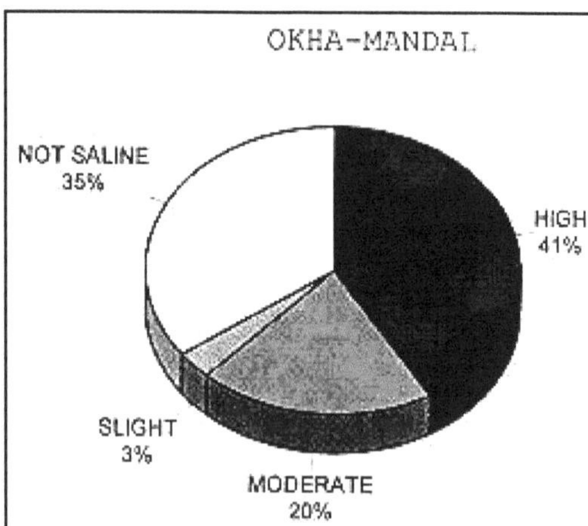

Figure 4.4: Pie diagram showing the distribution of salinity affected agricultural area.

Distribution of Area Affected by Salinity

Calculating area under different categories provide very useful insight for land development plan. Pie diagrams (Figure 4.4) show the distribution of area under various intensities of salinity for the taluka. The diagrams present the distribution according to the percentage of total agricultural land.

On an average 65 per cent of total agricultural land is affected by salinity in the taluka. Out of this, 42 per cent of agricultural area is highly affected by salinity, which covers 220.23 sq. km. area of the taluka. About 186.17 sq. km. of an area is comparatively better cultivated land and is not affected by salinity.

This makes about 35 per cent of agriculture area located in the middle part of the taluka. This region is geographically located far from coast and has remained unaffected by marine ingression.

Table 4.7: Distribution of agriculture area by degree of salinity

Percentage of Agricultural Area Falling Under Each Category	
Category	*Area in per cent*
Not Saline	35.08
Slightly Saline	3.23
Moderately Saline	20.2
Highly Saline	41.49

Source: Remote Sensing Imageries, SAC.

The above analysis indicates that the quality of agricultural areas in Okha Mandal taluka is deteriorated due to salinity ingression. This is one of the major reason contributing to the out migration of rural population from this taluka. In short, one can say that the decline and deterioration in agricultural areas play important role in development of the region.

Conclusion

The paper shows that one of the major problems of the Gujarat coastal region is salinity in soil and ground water, which has existed mainly as a result of marine ingression. The salinity affected agricultural areas cannot support a large number of populations. The carrying capacity of the land has declined due to the salinity ingression. There is a need to take this issue seriously in development planning.

The cropping pattern for the taluka which is dominated by ground nut and till cultivation needs an urgent diversification in order to preserve the fertility of the soil. Crops should be selected as per the degree of salinity present in the agricultural land. The cropping pattern suggested by agricultural department is listed in Table 4.8.

It is also important to improve the perception of environmental degradation and sustainability by adopting suitable land use pattern among the coastal community. Sustainable development cannot be achieved without active participation of local people. The role of NGOs cannot be overemphasized in this matter. The strong network

Table 4.8: Suggested cropping pattern as per degree of salinity

Degree of Salinity		
High Salinity	*Moderate Salinity*	*Slight Salinity*
Barley	Mustard	Field Beans
Beets	Wheat	Radish
Cotton	Juwar	Green Beans
Bermuda Grass	Tomato	Orange
Date	Lettuce	Grape Fruit
Asparagus	Carrots	Lemon
Spinach	Grapes	Plum
Turnip	Fig	Peach
Sugar beet	Olive	Apricot
Sugarcane	Potato	Pear
Tobacco	Sunflower	Peas
	Rice	Gram
	Maize	
	Cabbage	
	Onion	
	Banana	

Source: Dept. of Agriculture, Ahmadabad.

among government, non government organizations and local community can play important role in changing mind set for adopting suitable crops as per the land capability.

Methods and Case Studies

5

Spatial Decision Support Systems for Sustainable Land-Use Suitability Analysis: An Example from Ebro Basin in Zaragoza (Spain)

M.T. Lamelas, O. Marinoni, A. Hoppe & J. de la Riva

INTRODUCTION

The city of Zaragoza is located in the central part of the Ebro Basin (Figure 5.2). In this sector, the Quaternary alluvial terraces of the Ebro River, which were deposited above Tertiary gypsum formations, conform a covered karst area with intense karstification processes. These Quaternary materials are an important source of sand and gravels, utilized in the construction of new infrastructures, but also of groundwater, used for domestic, industrial and agricultural purposes.

The resources abundance was one of the reasons of the fast development of the city in the last decades. But this fast development brought in some cases negative interactions with the environment. Land subsidence caused the destruction of many infrastructures as well as water supply networks, means of transport and industrial buildings. Due to the unorganized urban development many infrastructures occupied the best soils traditionally occupied by the orchards and also destroyed important natural areas. Moreover, the proliferation of new industries increased contamination of the aquifer.

Nowadays, a well planned sustainable development is a must according to European law and requires a thorough decision making process. Therefore, as the land use has strong iterations with the geosphere, both geo-resources and geo-hazards should play a more important role in the land use decision process. Thus, a research project was initiated which aims at facilitating a sustainable land-use decision making with the use of Spatial Decision Support Systems (SDSS).

SDSS combine the benefits of spatial analysis provided by a Geographic Information System (GIS) and the skills of Decision Support Systems. Objective preferences can be taken into account for a great variety of criteria and objectives which are present in every land-use management. SDSS have therefore become an important tool to support the sustainable development of urban areas.

In this project, suitability maps for several land-uses (industrial, irrigated agriculture) are created with the use of the Analytical Hierarchy Process integrated in ArcGIS. They support land-use management under geoscientific aspects in a fast developing area in the surroundings of Zaragoza. A variety of criteria are introduced in the evaluation process like dolines susceptibility, groundwater vulnerability, agricultural capability of the soils and valuable natural areas protection among other.

Methodology

The GIS-based land-use suitability analysis has been applied in a wide variety of situations including ecological and geological approaches, suitability for agricultural activities, environmental impact assessment, site selection for several facilities and regional planning (Voogd, 1983; Janssen and Rietveld, 1990; ; Malczewski, 2004; Hoppe *et al.*, 2006; Lamelas, 2007; Lamelas *et al.*, 2008a,b,c; 2009).

Any planning process must focus on a mix of hard (objective) and soft (subjective) information. The former are derived from reported facts, quantitative estimates, and systematic opinion surveys. The soft information represents the opinions (preferences, priorities, judgments, etc.) of the interest groups and decision makers. The idea of combining the objective and subjective elements of the planning process in a computer based system lies at the core of the concept of SDSS (Malczewski, 2004).

SDSS can be defined as an interactive, computer-based system designed to support a user or group of users in achieving a higher effectiveness of decision making while solving a semi-structured spatial decision problem (Malczewski, 1999).

One can distinguish three major groups of approaches to GIS-based land-use suitability analysis according to Malczewski (2004): (i) computer-assisted overlay mapping, (ii) multicriteria evaluation methods, and (iii) AI (Artificial Intelligence-soft computing or geocomputation) methods.

The Multicriteria Decision Making (MCDM) procedures (or decision rules) define a relationship between the input maps and the output map. The procedures involve the utilization of geographical data, the decision maker's preferences and the manipulation of the data and preferences according to specified decision rules.

There exist in literature different attempts to classify MCDM according to several authors (Voogd, 1983; Vincke, 1986; Pereira and Duckstein, 1993; Malczewski, 1999).

However, most of them agree that decision rules can be classified into multiobjective and multiattribute decision making methods.

Multiattribute techniques are discrete methods because they assume that the number of alternatives (plans) is given explicitly, while in the multiobjective methods the alternatives must be generated. In this study, as the alternatives are already predefined, we are going to concentrate on multiattribute techniques (Voogd, 1983; Malczewski, 1999).

Additive decision rules are the best known and most widely used MADM (Multiattribute Decision Making) methods in GIS-based decision making. We concentrate on these techniques in which we can differentiate several types (Malczewski, 1999): Simple Additive Weighting (SAW), Utility Function Method, Analytical Hierarchy Process (AHP), ideal point and concordance methods.

In the context of land suitability analysis it is important to differentiate between the site selection problem and the site search problem. The aim of site selection analysis is to identify the best site for some activity given the set of potential (feasible) sites. The problem is to rank or rate the alternative sites based on their characteristics so that the best site can be identified. If there is not a pre-determined set of candidate sites, the problem is a site search analysis (Malczewski, 2004).

Despite the existence of diverse methodologies, multiple criteria decision making (MCDM) methods have certain aspects in common. Alternatives represent the different choices of action available to the decision maker. Multiple attributes represent the lowest level of decision criteria. Decision weights are assigned to the attributes. Usually, these weights are normalized to add up to one (Gilliams *et al.*, 2005).

Thus, several steps have to be covered in the suitability analysis (Figure 5.1). They are:

- ☆ Definition of alternatives: here alternatives are already defined.
- ☆ Definition of constraints: areas with land-use restrictions.
- ☆ Definition of variables: important factors in the decision process.
- ☆ Transformation into criteria: standardization of variables.
- ☆ Assignation of decision weights: in our case, with the use of Analytical Hierarchy Process.

In this contribution a site search analysis where every pixel represents one alternative is performed; the used pixel size is 20 x 20 m. The SAW method integrated in ArcGIS 9.1. (ESRI, 2005) is applied in order to create suitability maps for the location of new industrial and irrigated areas (Figure 5.1). In SAW methods, the decision maker directly assigns weights of "relative importance" to each attribute. A total score is then obtained for each alternative by multiplying the importance weight assigned for each attribute by the scale value given to the alternative on that attribute and summing the products over all attributes.

A variety of social, economical and environmental factors, playing an important role in the decision making process, are taken into consideration. But, previously, these variables or factors must be standardized in order to bring all of them to a common scale, criteria (Figure 5.1). According to Malczewski (1999), the classification (standardization) approaches can be classified in: Linear Scale Transformation, Value/utility Function, Probabilities, Revised Probabilities and Fuzzy Set Membership. In this project, the method used was the subjective scales or value function approach.

A fundamental problem of decision theory is how to derive weights of a set of criteria according to their importance. An excellent and well-known weight evaluation method is AHP proposed by Saaty (1977). This method, used in both approaches for criteria weights assignation, involves pair wise comparison to create a ratio matrix. Specifically, the weights are determined by normalizing the eigenvector associated with the maximum eigenvalue of the (reciprocal) ratio matrix. The procedure consists of three major steps: generation of the pair wise comparison matrix, the criterion weight computation, and the consistency ratio estimation (Saaty, 1977).

As mentioned above, all identified criteria are compared against each other in order to create a ratio matrix. Thus, numerical values expressing the preference of one

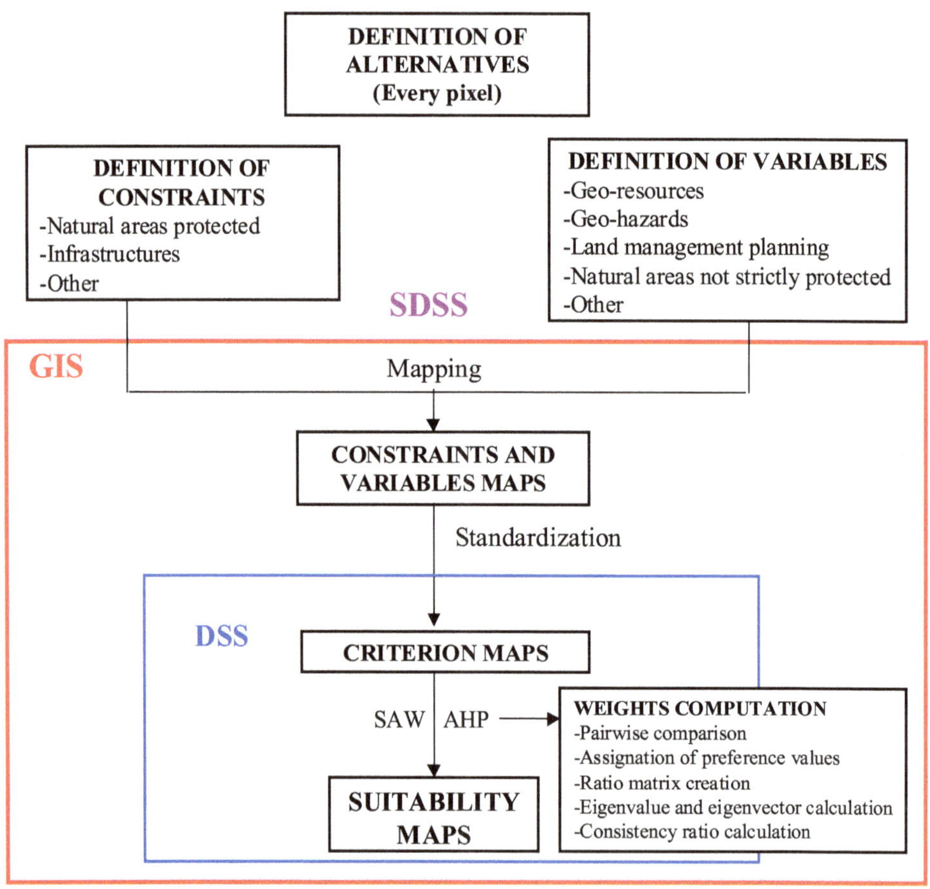

Figure 5.1: Workflow of the land-use suitability analysis.

criteria against another should be assigned. Saaty (1977) suggested a scale for comparison consisting of values ranging from 1 to 9; value 1 being equal importance and value 9, extreme prevalence (Table 5.1).

Afterwards, the assigned values are synthesized to determine a ranking of the criteria in terms of numerical values which are equivalent to the weights of the factors. Therefore, the eigenvalues and eigenvectors of the square preference matrix are calculated (Marinoni, 2004). According to Saaty & Vargas (1991), it is enough to calculate only the eigenvector resulting from the largest eigenvalue since this one contains sufficient information to determine the relative importance of the criteria.

Although the preference values are usually not arbitrarily assigned, the preference values selected by decision makers may be inconsistent and lead to perturbations in the eigenvector calculations. For this reason, Saaty (1977) additionally provided a single numerical index to check for consistency of the pair wise comparison matrix. The consistency ratio CR is the ratio of the consistency index CI to an average consistency index RI, thus

$$CR = CI/RI \quad (1)$$

The resulting average consistency index RI was calculated by Saaty (1977) as the average consistency of square matrices of various orders n which was filled with random entries. According to Saaty (1977), matrices with an order greater than 8 have a RI order of about 1.45. The consistency index CI can be directly calculated from the preference matrix with

$$CI = (\lambda_{max} - n)/n - 1 \quad (2)$$

where,

λ_{max} is the greatest eigenvalue of preference matrix and n the order of matrix. It is recommended that the consistency ratio presents values below 0.1.

Table 5.1: Scale for comparisons

Importance Value	*Description*
1	Equal importance
3	Moderate importance
5	Strong or essential importance
7	Very strong importance
9	Extreme importance
2, 4, 6, 8	Intermediate values
Reciprocals	Values for inverse comparison

Source: Saaty & Vargas, 1991.

The integration of the AHP in GIS combines decision support methodology with powerful visualization and mapping capabilities which in turn should considerably facilitate the creation of land-use suitability maps (Marinoni, 2004).

Suitability Analysis for Industrial Location

As mentioned above, in the site search approach every pixel in the criterion maps constitutes an alternative, thus alternatives are already defined.

Besides, some constraints representing the areas under extraction restrictions must be defined, these are:

- Natura 2000 network areas: natural reserve of the oxbows in La Cartuja, map provided by the Aragon Government.

- ☆ Urbanized areas: developed from the topographic map scale 1:25,000 from the National Geographical Institute (IGN).
- ☆ Infrastructures and their area of protection: roads, railroads, canals: also developed from the topographic maps.
- ☆ Other restrictive planning: Zaragoza land management planning, provided by Zaragoza Council, and natural resources planning, provided by the Aragon Government.
- ☆ Cattle tracks: seasonal migration of livestock, provided by the Aragon Government.
- ☆ Industrial areas where no space is left for new industries. Provided by the IAF (Instituto Aragonés de Fomento) from the Aragon Government.

A variety of social and environmental attributes or criteria are taken into consideration as well. Figure 5.2 shows the mapping of all the criteria: the less proper areas to industry location are in red while prone areas are in green. The criteria to introduce are:

- ☆ Natural environmentally important areas avoidance: SPAs (Special Protection Areas for birds), SACs (Special Conservation Areas), habitats, points of geological interest and other areas which mapping has been provided mainly by the Aragon Government.
- ☆ Dolines susceptibility: model developed with a logistic regression technique (Lamelas *et al.*, 2008).
- ☆ Groundwater protection: model developed by the German Geological Surveys (Hölting *et al.*, 1995; Lamelas *et al.*, 2007).
- ☆ Flooding hazard: cartography developed by Ollero (1996).
- ☆ Agricultural capability of the soils: mapping developed with the Cervatana Model (de la Rosa and Magaldi, 1982; Lamelas *et al.*, 2008b).
- ☆ Slope of the terrain: developed from the DEM (resolution 20x20 m) from the Ministry of Agriculture (SIG oleícola).
- ☆ Geo-technical characteristics of the subsoil: developed on the basis of the Geological map scale 1:50,000 from the National Geological Institute (ITGE).

To bring all criteria to a common scale, they must be standardized. The standardization method used was the Value Function approach according to Malczewski (2004), as the variables are classified into 6 categories: from 1 to 3 negative values and from 4 to 6 positive values.

Weights for criteria are assigned with the help of AHP where preference values have to be inserted into a square matrix where all criteria are pair wisely compared. Figure 5.3 shows this matrix within the windows of the tool developed for the GIS environment. These windows also show the weights assigned to every criteria when the "calculate" button is activated. In this case the consistency ratio (CR), which measures the consistency of the values assigned in the criterion matrix, presents a value of 0.021.

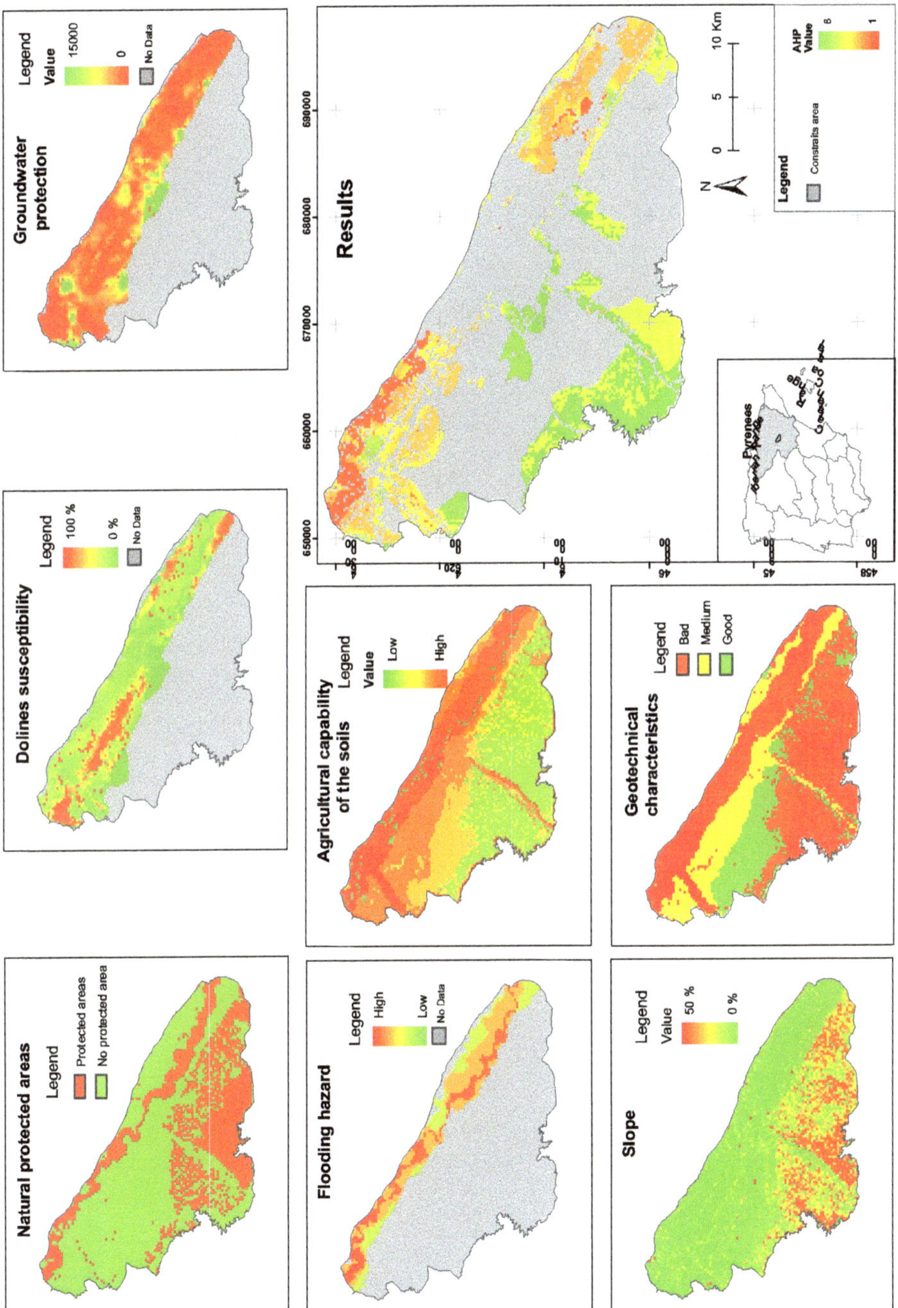

Figure 5.2: Results from the industrial use suitability analysis and original values.

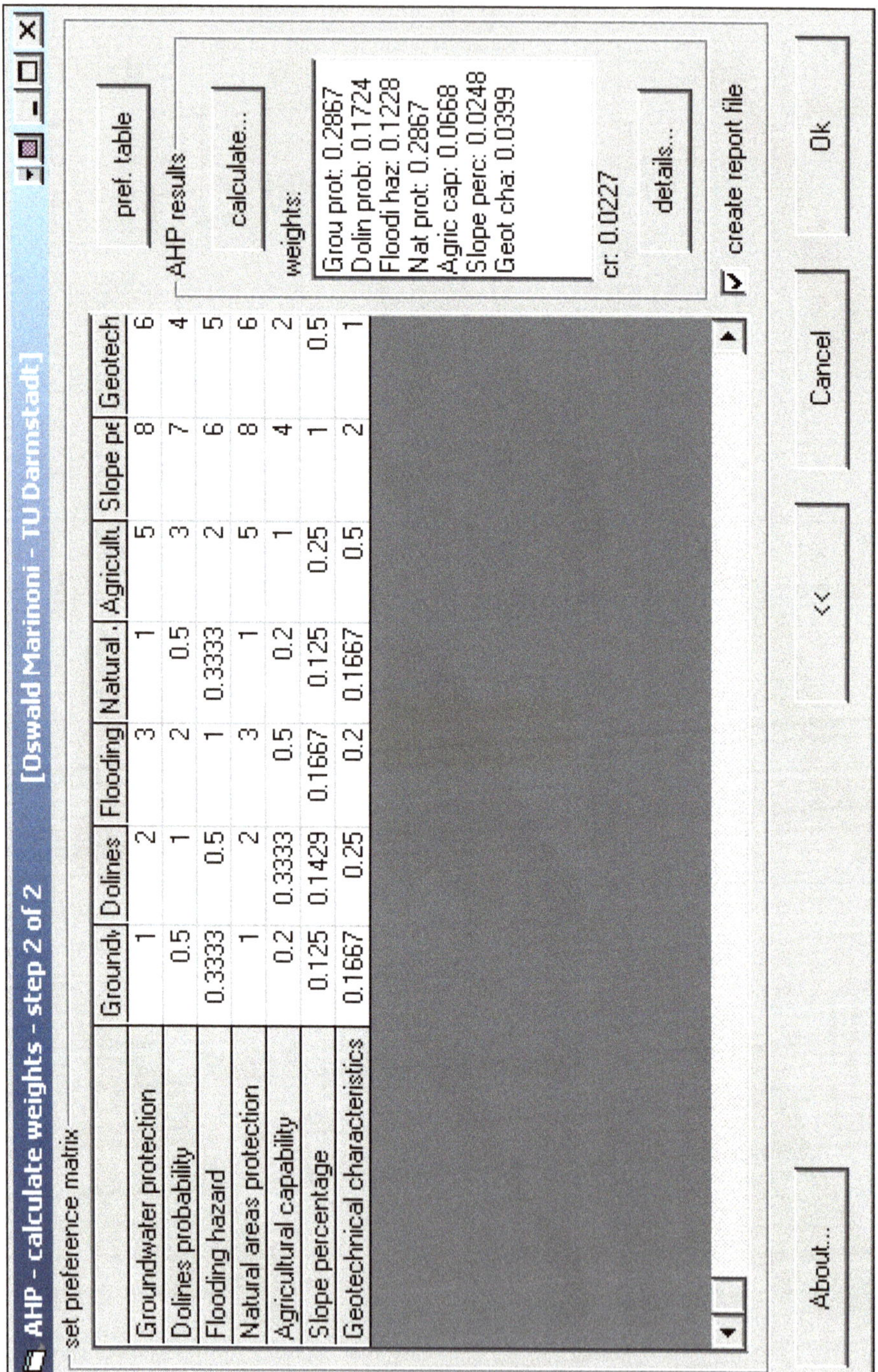

Figure 5.3: Pairwise comparison matrix, criteria weights and consistency ratio for suitability evaluation with respect to industrial locations.

The criteria preference values have been assigned by the authors of this paper after some talks with different stakeholders belonging to the Zaragoza Council and the Ebro River authority. The highest preference values are given to the ground water vulnerability and natural protected areas. This reflects our feeling that a sustainable development requires that the protection of groundwater and environmentally high valued areas should be given preference. Therefore these criteria obtained the highest weights, followed by the different hazards considered less important as this hazards can be avoided by application of new construction techniques as is the case of the last two factors in importance, slope and geo-technical characteristics of the soils.

Finally, all the classified raster files (criteria) are multiplied by its corresponding weight and summed up following a Weighted Linear Combination (WLC) approach, also known as Simple Additive Weighting (SAW) method.

Suitability Analysis for the Location of Irrigated Agricultural Areas

In the case of the irrigated agriculture location, the constraints are relatively similar, these are:

- ☆ Natura 2000 network areas
- ☆ Urbanized areas.
- ☆ Infrastructures and their area of protection: roads, railsroads, canals.
- ☆ Zaragoza land management planning and natural resources planning.
- ☆ Cattle tracks.
- ☆ Areas already irrigated. Information available at the Ebro River Authority web page (http://www.chebro.es).

In relation of the criteria these are relatively different, although with some factor in common (Figure 5.4), these are:

- ☆ Natural environmentally important areas avoidance: SPAs, SACs, habitats, points of geological interest and other areas.
- ☆ Dolines susceptibility.
- ☆ Groundwater protection.
- ☆ Erosion hazard: model developed with a qualitative weighing method developed by van Zuidam and van Zuidam-Cancelado (1979). The approach used here express the reduction in erosion hazard when irrigating (Lamelas *et al.*, 2008b).
- ☆ Agricultural capability of the soils to be irrigated: methodology based on the work performed by INARSA (1992).
- ☆ Slope percentage.
- ☆ Elevation above see level which limits the probability if irrigate under low costs. The source is the DEM from the Ministry of Agriculture.

The standardization method used is the same as the one used in the location of new industries. The variables are classified into 6 categories: from 1 to 3 negative values and from 4 to 6 positive values.

Weights for criteria were assigned also with the help of an AHP extension which was developed for the ArcGIS environment at the Institute of Applied Geosciences of the Darmstadt University of Technology. This tool can be downloaded from the ESRI web page (http://arcscripts.esri.com/). For more information about the development of this tool, the interested reader may refer to Marinoni (2004).

Figure 4.5 shows the matrix used for the irrigation location suitability analysis, the criteria weights and the consistency ratio. The weights have been assigned also by the authors of this paper, but in this case after some talks with stakeholders belonging to the Department of Irrigation and Soils from the Agricultural Research Center (CITA -Centro de Investigación y Tecnología Agroalimentaria), Aula Dei (CSIC-Scientific Research National Center) and Ebro River Authority.

We have assigned the highest weights again to groundwater vulnerability and natural protected areas for the same reasons as mentioned above. However, in this context it must be noted that the irrigation in this semi-arid region can be a source of contamination of the aquifer as the major recharge of the aquifer comes from irrigation returns. Dolines susceptibility is also an important factor as irrigation is considered as one of the main important factor to trigger doline development. Slope and elevation factors take the lowest values. The consistency ration is higher for the irrigation suitability analysis than for industrial location, 0.036; however, it is well below the recommended limit of 0.1.

Results

Figure 5.2 shows the final results of the land use suitability analysis for new industrial areas location. The best location for new industries is on the pediments and Tertiary materials outside of natural protected areas where the groundwater vulnerability and flood risk is lower. On the contrary, the worst location are the floodplains with high groundwater vulnerability values, also the natural protected areas around the river bed and other areas in the higher terraces which present more susceptibility to dolines development.

Figure 5.4 shows the final irrigation suitability map. The more suitable areas are located in a small sector in the south west of the study area. In this area, the pediments give a smooth slope to the terrain with low values of groundwater vulnerability and dolines susceptibility. The lowest values of suitability (red and orange values) correspond to the forest and thickets in the vicinity of the river and steppes in the south east of the area, which are protected (SACs and SPAs). In addition, the last ones present high slope values.

Conclusions

According to Agenda 21, countries should undertake efforts to build up a comprehensive national inventory of their land resources in order to establish land information systems. These systems will help to classify land resources according to their most appropriate uses. Moreover, environmentally fragile or disaster-prone areas can be identified better for special protection measures.

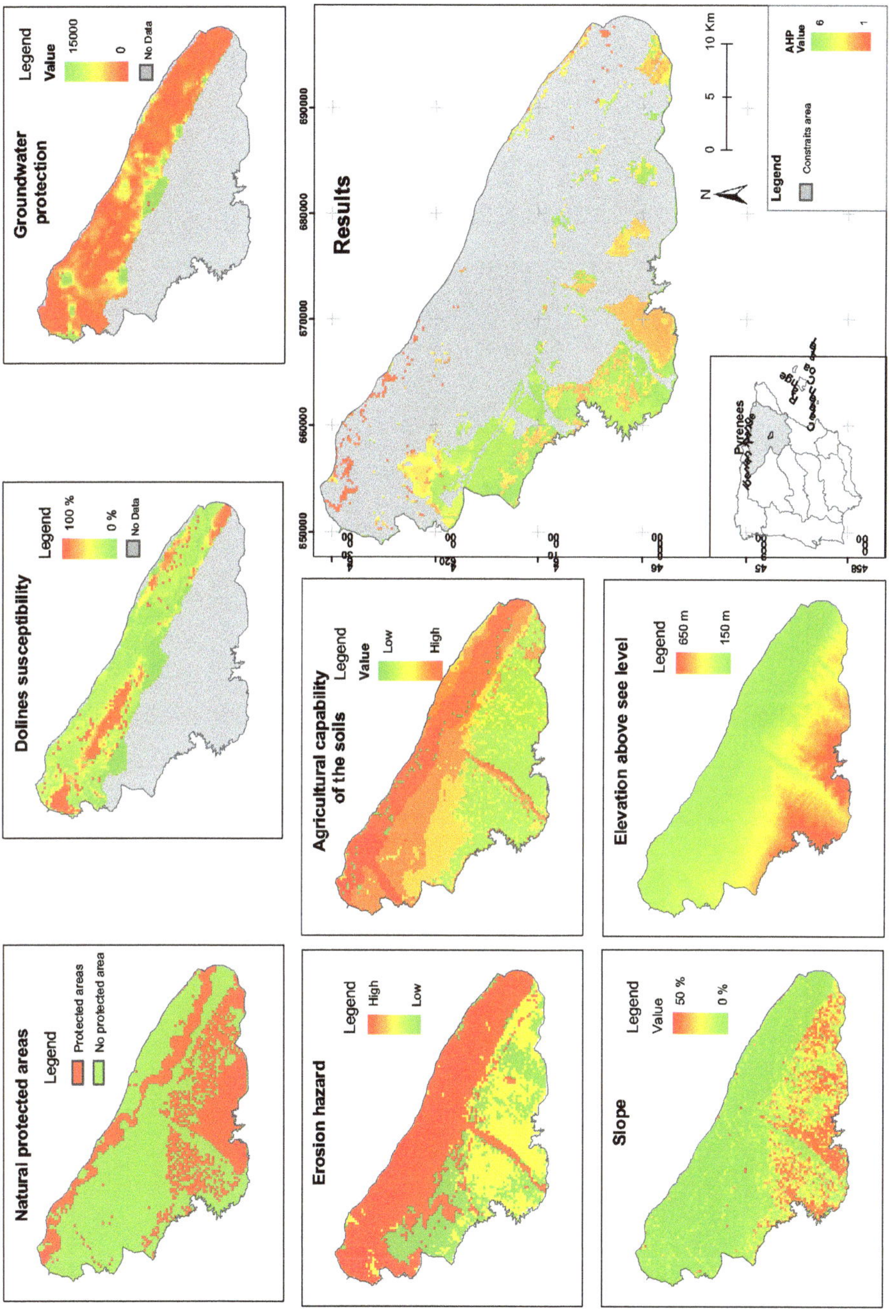

Figure 5.4: Irrigated agricultural use suitability.

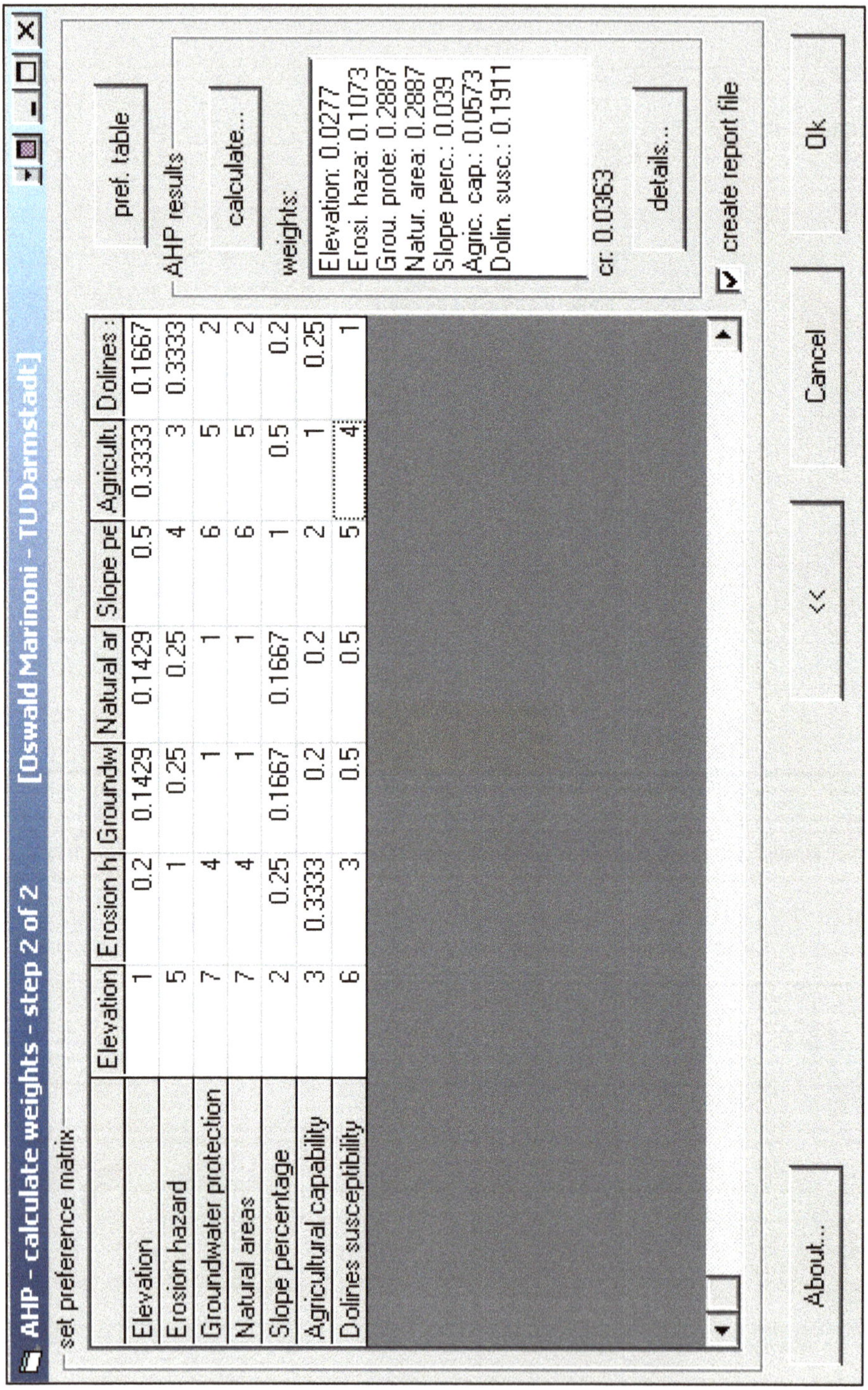

Figure 5.5: Pairwise comparison matrix, criterion weights and consistency ratio for irrigated agricultural use suitability.

The overall objective is to improve or restructure the decision-making process so that consideration of socio-economic and environmental issues is fully integrated and a broader range of public participation is assured.

In the last years, the development of SDSS has been a considerable aid in the efforts for solving land-use conflicts that commonly appear in a sustainable land-use management.

The greatest disadvantage of the SAW methods is that they tend to be ad hoc procedures with little theoretical foundation to support them. However, since they are easy to use, SAW methods are actually quite widely applied in real-world settings.

A fundamental problem of decision theory is how to derive weights of criteria. A well-known weight evaluation method is the AHP. However, one disadvantage of this method is the inherent subjectivity of assigning preference values between criteria.

Nevertheless, it should not be forgotten that land-use decisions are made by managers, implying a certain level of subjectivity.

In the surroundings of Zaragoza, which represent a fast developing area, the integration of SAW and AHP methods in a GIS has been a substantial aid in the search for future industrial and irrigated agriculture locations. We would like to stress the value associated with the integration of geoscientific aspects in the land use decision process as many geo-resources are rare and valuable or can be only used once.

Acknowledgements

This research is funded by the Deutsche Forschungsgemeinschaft (DFG, Ho 804/7-1+2). Thanks to José Ángel Navamuel (Zaragoza Council), Miguel Ángel García Vera, José Losada (Confederación Hidrográfica del Ebro), José Albiac and Daniel Isidoro (CITA) and Enrique Playán (Aula Dei) for their useful advise.

References

de la Rosa, D. and Magaldi, D. 1982. Rasgos metodológicos de un sistema de evaluación de tierras para regiones mediterráneas. Sociedad Española de la Ciencia del Suelo, Madrid.

ESRI Inc. 2005. Arc GIS 9.1. ESRI Inc., Redlands, CA.

Gilliams, S., Raymaekers, D., Muys, B. and van Orshoven, J. 2005. Comparing multiple criteria decision methods to extend a geographical information system on afforestation. *Computers and Electronics in Agriculture* **49:**142-158.

Hölting, B., Haertlé, T., Hohberger, K.-H., Nachtigall, K.H., Villinger, E., Weinzierl, W. and Wrobel, J.-P. 1995. Konzept zur Ermittlung der Schutzfunktion der Grundwasserüberdeckung.*Geologisches Jahrbuch* **C 63**:, 5-24.

Hoppe, A., Lang, S., Lerch, C. and Marinoni, O. 2006. Geology and a spatial decision support system for the surroundings of urban areas: An example from southern Hesse (Germany). *Zeitschrift der Deutschen Gesellschaft für Geowissenschaften* **157**:135–146.

INARSA 1992. Estudio de la viabilidad técnico-económica del canal de la margen derecha del Ebro y sus alternativas en Aragón. Diputación General de Aragón, Zaragoza.

Janssen, R. and Rietveld, P. 1990. Multicriteria analysis and GIS: An Application to Agricultural Land-use in the Netherlands. In Scholten, H. and Stillwell, J. (Eds.), *Geographical Information Systems for Urban and Regional Planning.* Kluwer Academic Publishers, Dordrecht,

Lamelas, M.T. 2007. *Geo-resources and Geo-hazards in the Context of a Sustainable Development in the Periphery of Urban Areas, Exemplary of a Part of the Ebro Basin in the Surroundings of Zaragoza (Spain).* Ph.D., Technische Universität Darmstadt, Darmstadt. P211. Available on http://elib.tu-darmstadt.de/diss/000794, last accessed April, 2010.

Lamelas, M.T., Marinoni, O., Hoppe, A. and de la Riva, J. 2007. Groundwater vulnerability map for the Ebro alluvial aquifer between Jalón and Ginel tributaries (Spain). *Environmental Geology* **53:**861-878.

Lamelas, M.T., Marinoni, O., Hoppe, A. and de la Riva, J. 2008a. Dolines probability map using logistic regression and GIS technology in the central Ebro Basin (Spain). *Environmental Geology* **54:**963-977.

Lamelas, M.T., Marinoni, O., Hoppe, A. and de la Riva, J. 2008b. Suitability analysis for sand and gravel extraction site location in the context of a sustainable development in the surroundings of Zaragoza (Spain). *Environmental Geology* **55**:1673–1686.

Lamelas, M.T., Marinoni, O., Hoppe, A. and de la Riva, J. 2008c. Modelling environmental variables for geohazards and georesources assessment to support sustainable land-use decisions in Zaragoza (Spain). *Geomorphology* **111:**88-103.

Malczewski, J. 1999. *GIS and Multicritera decision analysis,* Wiley, New York, p 392.

Malczewski, J. 2004. GIS-based land-use suitability analysis: a critical overview. *Progress in Planning* **62:** 3-65.

Marinoni, O. 2004. Implementation of the analytical hierarchy process with VBA in Arc GIS. *Computers and Geosciences* **30:**637-646.

Marinoni, O. 2005. A stochastic spatial decision support system based on PROMETHEE. *International Journal of Geographical Information Science* **19:**51-68.

Ollero, A. 1996. *El curso medio del Ebro,* Consejo de Protección de la Naturaleza de Aragón, Zaragoza, p 311.

Pereira, J.M.C. and Duckstein, L. 1993. A multiple criteria decision-making approach to GIS-based land suitability evaluation. *International Journal of Geographical Information Systems* **7:**407-424.

Saaty, T.L. 1977. A scaling method for priorities in hierarchical structures. *Journal of Mathematical Psychology*. **15**:231-281.

Saaty, T.L. and Vargas, L.G. 1991. *Prediction, Projection and Forecasting,* Kluwer Academic Publishers, Dordrecht, p 251.

van Zuidam, R.A. and van Zuidam-Cancelado, F.I. 1979. *Terrain analysis and classification using aerial photographs. A geomorphological approach, ITC textbook of photo-Interpretation, 7,* ITC. The Netherlands. p 310.

Vincke, P. 1986. Analysis of multicriteria decision aid in Europe., *European Journal of Operational Research* **25:** 160-168.

Voogd, H. 1983. *Multicriteria Evaluation for Urban and Regional Planning.* Pion Limited, London, p 367.

6

Land-Use and Land-Cover Change: Case of Jhansi in Bundelkhand

R.S. Khoiyangbam & Santaram S. Oinam

INTRODUCTION

Land resource is the basic and important resource after human population (Gurjar and Prasad, 2007). The term land-use refers to how the land is being used by human beings and land-cover refers to the biophysical materials found on the land (Jensen, 2000). The land-use pattern of an area is directly related with the level of techno-economic advancement (Whyte, 1961). Land-use is a dynamic phenomenon, and both its value and pattern changes from one particular point of time to another and also from geographical unit to another, with varying efficiencies, abilities, priority and needs (Bisht and Tiwari, 1996). The information on land-use/land-cover patterns, their spatial distribution and changes over a time scale are the prerequisite for making developing plans (Dhinwa *et al.*, 1992). Land-cover and land-use data present information about the physical coverage of the earth surface and its socio-economic purpose. As such, land-cover and land-use information can be regarded as basic information, describing the composition and the configuration of the territory. In view of the pressure exerted by increasing demand of land resource, appropriate scientific land-use planning and land management strategies could provide, the alternative for the sustainable development of any region (Saxena, *et al.*, 1990). Land-cover and land-use can be seen as the interface between natural conditions and human influence providing a sound framework for linking socio-economic development with some environmental impacts. Of specific interest is the analysis of changes in land-cover and land-use as a consequence of socio-economic development and its impacts on the health and environment. Land-use management has been studied by various authors in India (Kuniyal, *et al.*, 2009; Kala, *et al.*, 2008; Oinam, *et al.*, 2007, Oinam, *et al.*, 2005). The objective of this paper is to analyze from the data available, the general pattern of distribution of educational land among its important uses in institutes. The finding of the study will benefit educationalist, scientist, technocrats and planners to find out possibilities of securing a more sustainable way of managing and utilization the institutional land.

The Study Site–Jhansi District

The district of Jhansi lies in the southwest portion of Jhansi division of Uttar Pradesh state of India between 25° 30′ and 25° 57′ North latitude and 78° 40′ and 79° 25′ East longitude. The present area of the district, according to survey of India is 5,024 km^2, which comprises of four tehsils, *viz.* Garautha (1564 km^2), Moth (1193 km^2), Jhansi (1192 km^2) and Mauranipur (1078 km^2). Climatically Jhansi falls under a semi-arid climate, with two main seasons: monsoon and dry. The monsoon brings over 90 per cent of the annual rainfall between the months of June to September. Peak summer (May-June) brings excessively high temperatures, often topping 40° C, as the hot, dry loo winds sweep in from the desert. Indeed, the area is notorious for experiencing droughts in summer. During the winter months (December-February) daytime temperatures are quite pleasant reaching highs between 16.5 and 21° C. Nighttime brings much cooler temperatures and frost has been known to occur on the coldest evenings. Soil in the Jhansi is often divided into two major types: red and black. Red soils are more predominant in the northwestern region and are usually gravel and shallow with consequently poor moisture retention. Conversely, black soils, which are found mostly in the south, are more water retentive and are therefore preferred for wheat, gram and sugarcane cultivation. Covering only a small percentage of the Jhansi, riverine soils are primarily formed by sand and clays.

Methodology

As the demand for land – a finite resource is increasing, it is necessary to make a detailed survey of land availability in each village and town. Changes in land-cover and use can be studied by means of various approaches and at several levels, depending on the available information. Simple area balances approach presents the first level and the simplest way of analyzing changes in land-cover (land-use). Changes in the area estimates, widely published in many census records, statistical yearbooks, etc., are the core sets of area statistics, which can be conveniently employed. These data provide quantitative information of changes in land-use and its dynamic processes giving an overview of the type of human influence (economic purpose and landscape utilization). However, it reveals limited conclusions on the quality or their environmental and human health impacts. A certain land-cover category may be stable in its extent, but its "ecological quality" may, nevertheless, be deteriorating or improving. Only if qualitative aspects are reflected in the land-cover classes, environmental statements are possible. In the present study primary and secondary data pertaining to the educational institute land-use and land-cover was colleted from various offices in the district. The data were used to analyze and interpret the changes in institutional land-use and land-cover in the district. Visits were made to different primary, high and intermediate schools and also to general and technical colleges, university and research centers located in the district to observe the pattern of land utilization within the four walls of the institute. Qualitative aspects regarding the probable impacts on the environment and human health due to institutional land-use pattern was determined.

Results and Discussion

General Land-Use Pattern in Jhansi

The total geographical area of the district Jhansi in 2003-04 was around 4,99,393 hectares. The detail of land utilization in Jhansi in 2003-04 is depicted in Figure 6.1. The net sown area was 3,26,079 ha, which accounts to about 65.19 per cent total land and was the largest form of land-use type in the district. The land under the present fallow (42,580 ha), land put to non agricultural use (41,334 ha), forest land (33,638 ha) and Barren and uncultivated land (31104 ha) occupies 8.53 per cent, 8.28 per cent, 6.74 per cent and 6.23 per cent, respectively of the total area and were the four important land-cover after the cultivated land. Other uses of land, such as towns and village settlements, roads and bridle-paths, canals, lands under the cover of educational institutes, schools, colleges, university, *etc.*, although occupy relatively small areas is

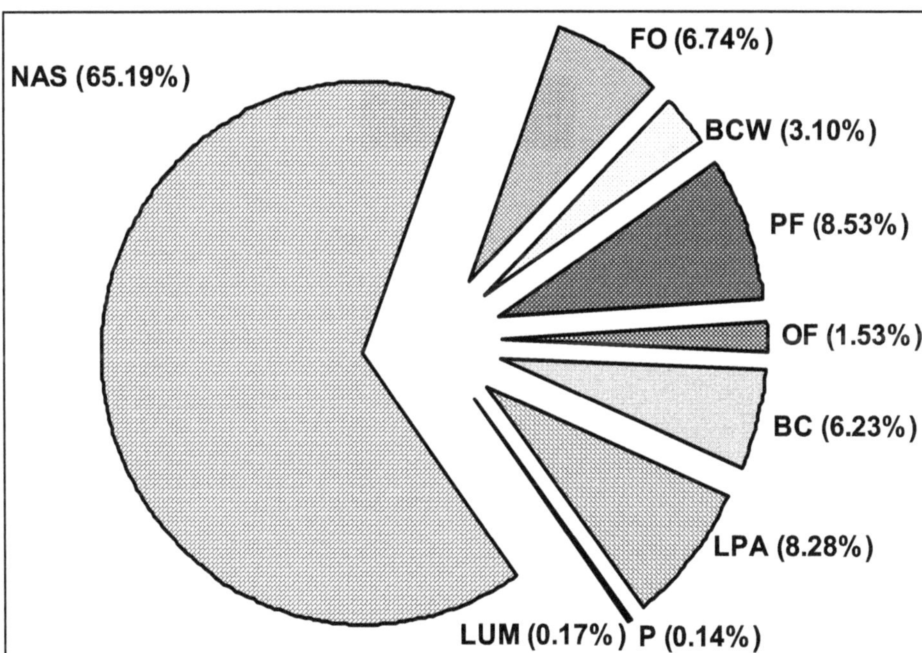

Figure 6.1: Land-use pattern in Jhansi in 2003-04.

NAS: Net area sown; **FO**: Forest land; **BCW**: Barren and uncultivated land; **PF**: Present fallow land; **OF**: Other fallow land; **BC**: Barren and uncultivated land; **LPA**: Land put to non agricultural use; **P**: Pasture; **LUM**: Land under miscellaneous trees and grove.

very important from the economic standpoint. The land-cover under the pastures (677 ha) and under miscellaneous trees and grove (829 ha) were less than 2 per cent of the land area.

Land-Use Changes in Jhansi

The land-use change that took place in the district within the period of 1991-92 and 2003-04 are given in Table 6.1. The net sown area has increased by 15,367 ha from 310712 ha (1991-92) to 326079 ha (2003-04). Such an increase in the area of land under crop has been achieved partly by bringing new areas under the plough and by more intensive use of the existing land. Increasing pressure of population on the land and the need to increase food production has resulted in increase of cropped area. Within the period the land under forest (33,274 ha), present fallow land (22,828 ha), land put to non agricultural use (40,093 ha) and pastures (1983 ha) had also shown a recorded expansion by 864, 19,752, 1,241 and 34 ha, respectively. The expansions were mainly achieved largely by conversion of the barren cultivated wasteland and partly from encroachment towards the other fallow land, barren and uncultivable land and land under miscellaneous trees and groves. The land area under the barren cultivable wasteland, other fallow land, barren and uncultivable land, land under miscellaneous trees and groves in 1991-92 was 47,779, 14,516, 31,514, 1,983 ha. A loss of 32,289, 6,854, 410 and 1,154 ha respectively, from the four land-use categories was recorded by 2003-04.

Table 6.1: Land-use changes in Jhansi district, Uttar Pradesh

Land-Use Types	*Area (hectares)*		*Changes*	
	1991-92	*2003-04*		
Forest	32774 (6.52)*	33638 (6.74)*	+	864
Barren cultivable wasteland	47779 (9.50)	15490 (3.10)	–	32289
Present fallow land	22828 (4.54)	42580 (8.53)	+	19752
Other fallow land	14516 (2.89)	7662 (1.53)	–	6854
Barren and uncultivable land	31514 (6.27)	31104 (6.23)	–	410
Land put to non agricultural use	40093 (7.97)	41334 (8.28)	+	1241
Pasture	643 (0.13)	677 (0.14)	+	34
Land under misc. trees and grove	1983 (0.39)	829 (0.17)	–	1154
Net area sown	310712 (61.79)	326079 (65.19)	+	15367
Total area	**502842 (100.00)**	**499393 (100.00)**	–	**3449**

* Figures in the parenthesis give the percentage of the land to that of the total area

Competition of Educational Land-Use with Other Land-Use Types

In Jhansi like many other parts of the country, as available land becomes limited, educational land-use often competes with other forms of land-use. In many cases lands which might have been otherwise purposefully used are converted into educational land due to various underlying compelling factors. In the urban areas the competition is more intense with land-use options such as manufacturing industries, commercial areas, settlement areas, offices, transport facilities like road and railways lines, *etc.* Whereas, in the rural areas the institutional land-use competes with land requirements largely for agricultural purposes. For the advantage of location and road linkages many a time good agricultural lands have to be compromise in favor of institutional development. Human and economic factors are equally important in deciding the institutional land-use in a region. Economic factor may also lead to improper land-use overlooking the capability of land. In densely populated areas like the Jhansi district the pressure of growing populations (4.27 lakhs in 1901 to 14.30 lakhs in 1991; Figure 6.2) has operated to bring under use all land that is usable and the pattern of land utilization comes much closer to the physical capability of the land. Hence, any change in use of land may have various environmental implications and therefore such change should be carried out only after developing a proper planning and management strategy. There is a need of overall zonation of land taking in consideration the land-use capability, environmental aspects, socio-economic and location. The vertical expansion of educational land-use through multi-storey buildings have upto some extent restricted the horizontal expansion in recent times. However, even in many such space efficiency institutes the minimum land requirements of playground, garden and a healthy environment are found to be grossly compromised for want of sufficient land.

Trends in Educational Expansion

The rapid expansion of population in the Jhansi district (Figure 6.2) coupled with the economic development and the efforts made by the governmental agencies to impart universal elementary education have resulted into the growth in the number of school going children. Concurrently, catering to the larger need various educational institutes have mushroomed up in the last few decades, private stakeholders taking

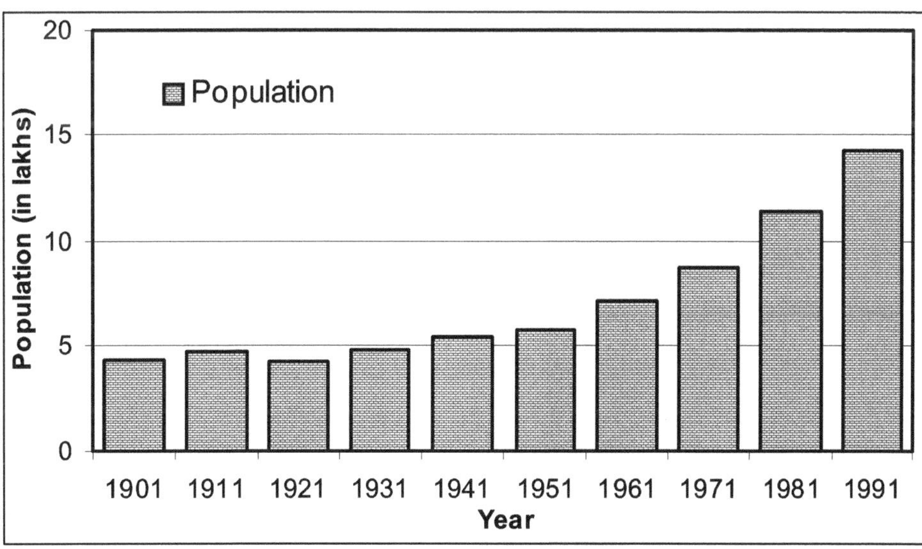

Figure 6.2: Population of Jhansi district.

an increasing role. During the period of three decades (1980-81 to 2002-03) the total number of students enrolled in the educational institutes (excluding the enrolment in university) increased from 1.79 to 4.07 lakhs (Figure 6.3). The distribution of students in the various educational imparting levels in 1980-81 was 1.03 lakhs (Junior basic schools), 0.32 lakhs (Senior basic schools), 0.33 lakhs (High/intermediate schools) and 0.11 lakhs (colleges), respectively. Similar distribution of students in the four different educational levels in 2002-03 was 2.05, 0.53, 1.24 and 0.25 lakhs. These corresponds to an increased by 199.03 per cent, 165.63 per cent, 375.76 per cent and 227.27 per cent. During the period of 1980-81 to 2002-03 the number of education institutes in the area increased from 1070 to 1535, equivalent to growth of 143.46 per cent (Figure 6.4). To a large extent educational institutes are classified on the basis of the levels and types of educations imparted. The number of Junior basic schools, Senior basic schools, High/intermediate schools and colleges increased from 103, 32, 33 and 11 in 1980-81 to 205, 53, 124 and 25, respectively in 2002-03.

Educational Land-Use and Land-Cover

In 1980-81 the educational institutional land-cover in Jhansi district was around 4216 hectares, which constitutes about 00.84 per cent of the total geographical area of the district. These land area expanded upto 6565 in 2002-03 constituting 1.31 per cent of the total land. These correspond to an increased by 155.72 per cent during the last two decades. These expansions even though small in area are very significant, because educational institutes tend to seek land in and around the settlement areas. Consequently, the expansion occurred more rapidly and disproportionately in the urban area compared to rural areas. However, private higher education institutions which have come up in large numbers since the early 1980s out numbering the governmental institutions are largely situated in the urban rural fringes. This might be due to reason that these institutions imparting technical education, demands more extensive open space. The rural urban fringe zone is formed where the rural and urban characteristics are mixed together. It is a zone of mixed land-use elements and characteristics in which rural activities and modes of life are in rapid retreat and into which not, only residential but also commercial educational, recreational, public service and other largely extensive users of land are intruding. The area is only partially assimilated into the growing urban complex. Most of the technical colleges and other institutes sprang up in this region and slowly engulfed in the expanding urban area. The institutions such as the Bundelkhand University and Jhansi Rani Medical college which currently comes under the urban area was then started earlier in the urban rural fringe before two decades. Many institutes of Engineering, pharmacy, research centers including the Indian Grassland Research Institute and the National research Centre for Agroforestry are at present located in the outer fringe of the city.

The results of the detailed survey conducted to reveal the institutional land-cover in 15 selected schools in Jhansi city is presented in Table 6.2. The land area of the schools ranged between 700 ft^2 in small schools (A. J. Memorial Public School) to 15000 ft^2 (Laxmi Vyayam Mandir Saraswati Balika). It is very true that the larger the land area the better learning and healthy environment can be expected. However, larger land area does not directly imply greater facilities and congenial environment

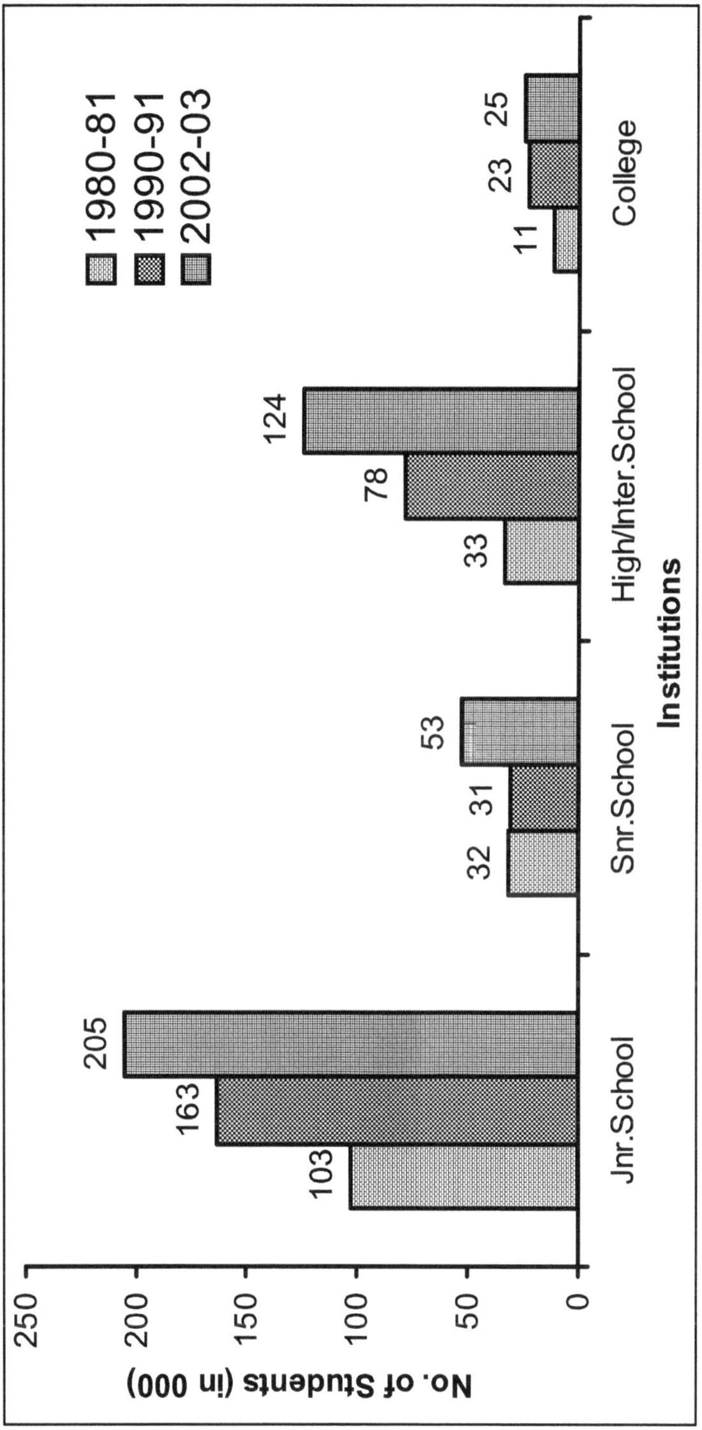

Figure 6.3: Students strength in different educational levels in Jhansi district.

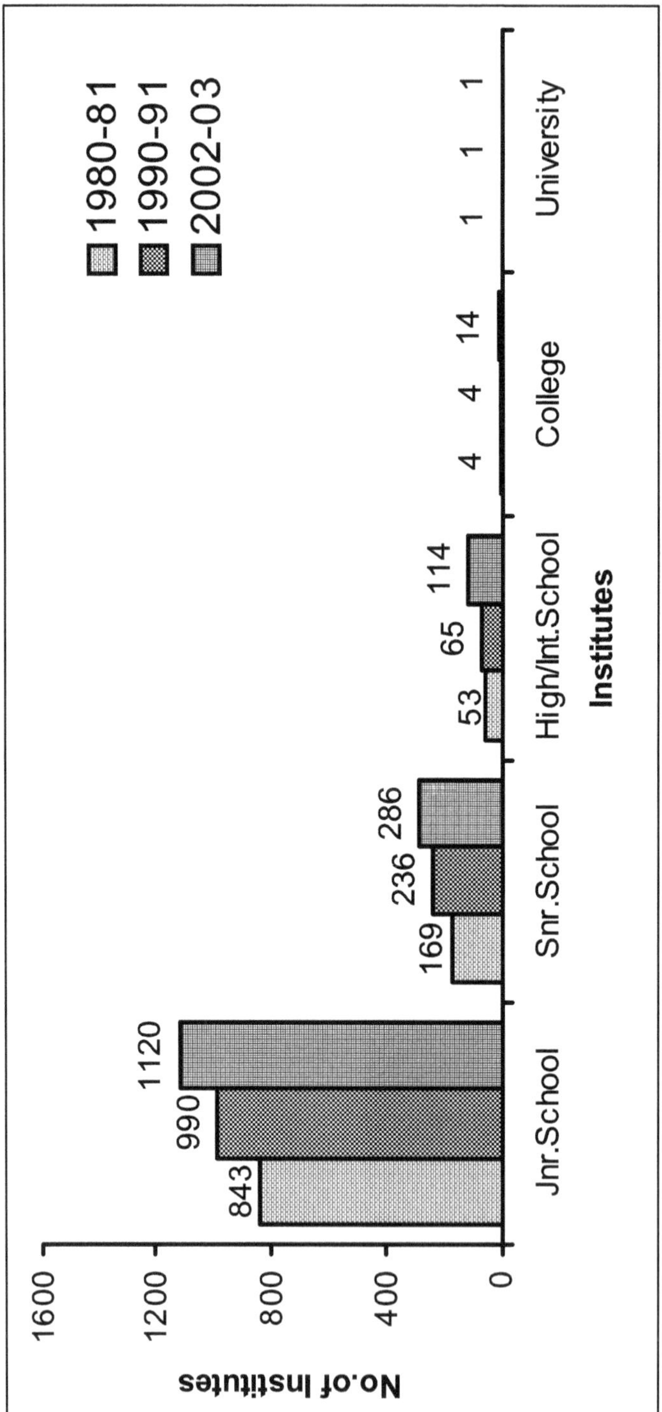

Figure 6.4: Number of Institutes in Jhansi district.

since many a times in many schools the available lands could not be properly and efficiently utilized. The size, shape and number of the institutional rooms vary considerably from school to school. Recreational spaces within the school for the children are either totally neglected or insufficient or poorly maintained if available. This may be partly due to lack of land as prime factor plus the shortages of economic resources. Out of the 15 schools under study flowering gardens were found in 10 schools and playing ground in 13 schools. Two schools have swimming pool facility for the children. However, in most of the schools this type of land-use was not given priority. The reason for non-existence of laboratory in some schools were due to level of education being imparted, in these schools the classes taught was only upto primary schools. The absence of library facilities in many of the schools might be due to the unavailability of sufficient space besides other reasons. Overall, on environmental and health point of view the schools under study were not encouraging, except for a few schools. In most of the schools the rooms were congested and were not well ventilated. Some of the rooms were devoid of both natural light and free aeration. In a worst but rare case, the exit door of one classroom opened up directly in another classroom. However, institutes such as Gyan Sthaly Public school, Shri Sinatan Dharma Kanya College, Laxmi Vyayam Mandir had well ventilated rooms. Most of the rooms in these institutions had more than two windows and wide ventilators aloof the roof and most of them accessed to natural light.

Table 6.2: Land-use and land-cover within the schools

Parameter/Facilities	*Schools with the facility*	*Ranges*		*Average**
		Min.	*Max.*	
Area (in 000 ft^2)	15	0.7	15	3.98
Class rooms (Nos.)	15	6	35	16.67
Laboratory rooms (Nos.)	12	0	8	3.13
Library rooms (Nos.)	3	0	1	0.20
Office rooms (Nos.)	15	1	2	1.33
Rooms (total Nos.)	15	9	46	21.33
Teaching staffs (individuals)	15	8	31	30.93
Doors classroom^{-1} (Nos.)	15	1	2	1.80
Windows classroom^{-1} (Nos.)	15	0	6	2.2
Swimming pool (Nos.)	2	0	1	13.30
Flower gardens	10	0	1	66.60
Playing ground	13	0	1	86.60
Cycle stand	7	0	1	46.60
Students (000 individuals)	15	0.1	3	988

*: Average of the 15 schools under survey

Educational Catchment Areas

The actual expansion of educational land-use is not restricted within the premises where the institutional buildings are housed but covers the whole catchment where the student and staffs are scattered beyond the office hours and where they reside beyond the institutional premises. For instance in case of Bundelkhand University, the highest seat of learning in Jhansi and adjoining districts, students, staff and faculties are drawn from a fairly wide expanse, touching every nook and corner of the country. Most of the individuals coming from far off distance reside in the nearby localities adjoining the University in rented houses. This population, in thousands of strength had certainly played a subtle role in shaping the land-use pattern and even the valuation of the land where they stay. Have there not been the Medical College and the Bundelkhand University established, the land surrounding these institutions might have settlement colonies developed at a much slower pace and the rented houses might have been considerably cheaper. Similarly, in many of the colleges established to impart technologies the students and staffs are coming from far off places and these led to an increase in complexity in determining the educational land-use changes.

Health Implications Associated with Educational Land-Use

The health and environmental problems associated with educational land-use may be partly due to its location and faulty design of institutional buildings. Institutes are located at places which have certain advantages. Many schools mean for young children in Jhansi city are located near busy streets, National highway and some in the commercial areas. Besides, the health risks associated with the road accident, the children are continuously exposed to the hazardous air pollutants and high intensity noise through out the schooling hours. Faulty design of the institutional buildings both due to lack of proper financial investment and due to lack of proper land area often led to poor ventilation and improper access to natural light. These may result into build up of indoor air pollutants exceeding the permissible level in the institutions. As institutional rooms have a larger number of occupants, large amounts of outdoor air must move freely through the building to assure proper aeration. Above all, the lack of land for playground, gardens and recreational space in the schools may indirectly affect the learning process and in delivering quality education.

Factors Affecting Institutional Land-Use Decision

Human and economic factors are equally important in deciding the land-use in a region. Economic factor may also lead to improper use land overlooking the capability of land. For *e.g.* good agricultural land along the margin of a large city may be converted for locating institutions. Other factors affecting the location of institutes and influencing institutional land-use pattern include, availability of land, ownership of the land, distance from the main town, characteristic of the soil etc. Taking in consideration of these factors institutes are located at places, which have certain advantages. The main factors of location in primary school is the students population, distance from the village, available space etc., whereas in terms of higher institutes and research centers power supply, transport are taken into account. Economic and political also influence institutional locations.

Need of Land-Use Planning in Educational Institutes

The pattern of land-use in the country as a whole and the changes that have taken place since independence indicate that land capability has not been taken into account. Marginal lands have been cultivated and this may result in permanent degradation of land. The ideal procedure would be to make an assessment of the land capability of all land areas and plan their use accordingly. Proper land-use planning would lead to better use of land with reference to our needs and aspirations. Any improper use of land leads to degradation of land and make it unsuitable for any use. Such improper use of land can be avoided by proper land-use planning. While devising and planning strategy for development of educational land besides other parameters, due consideration must be given both on the land capability, land availability and the land planning to efficiently utilized the institutional space without compromising health and environmental aspects.

Conclusions

Like other pattern of land-use, area under educational land-cover is expanding rapidly in rural, urban and rural-urban inter-fringes. Rapid population growth in the last 30 years has resulted in a high percentage of children in our population. As the country prepares to reach the goal of providing free and compulsory education to all the children, the demand of land for institutional purpose will also expand considerably. Most of the educational institutes in Jhansi are over congested. Planned utilization of land for educational purpose will be of great importance to economic, social and educational development and well being of the people in the district. The ideal procedure would be to make an assessment of the land capability of all land areas and plan their use accordingly. Proper land-use planning would lead to better use of land with reference to our needs and aspirations. Development of educational land must also give due importance to health and environmental aspects to attained sustainable land-use system in the district.

References

Anonymous (1966) Report of the Education. Education for National Development. Ministry of Education, Govt. of India, New Delhi.

Bisht, B. S. and Tiwari, P. C. (1996) Land-use planning for sustainable resources development in Kumaun lesser Himalaya – A case study of the Gomti watershed. *International Journal of Sustainable development and world Ecology.* **3:** 23-34

Dhinwa, P. S., Pathak, S. K., Sastry, S. V. C., Rao Mukund, Majumdar, K. L., Chotani, M. L., Singh, P. J. and Sinha, R. L. P. (1992) Land-use changes analysis of Bharatpur district using GIS. *Journal of the Indian Society of Remote Sensing.* **20(4):** 237-250

Gurjar, R. K. and Prasad, R. (2007) Land-use pattern of Sahibi River Basin – A Geographical study – Alwar District, Rajasthan. *Journal of Water and land-use management.* **7(2):** 181-188

Hargerstrand, T. (1995) Remote sensing, GIS and the landscape mantle, Systema Terra- *Remote Sensing and the Earth.* **4(2):** 7-10

Jagdish C. Kuniyal, Santaram S. Oinam, Yaswant S. Rawat, S.C.R. Vishvakarma and Dinesh C. Pandey (2009). Impact assessment of agricultural transformation on the land-use pattern and traditional agro-ecosystem in high altitude region of the Lahual valley, Northwester Himalaya. *In Management strategies for the Indian Himalaya: Development and Conservation, Vol. II.*, Eds. Rawat, M. S. S. and Pratap. D. Transmedia Publication Srinagar (Garhwal) Uttrakhand. Pp. 72-90

Jensen, J. R. (2000) Remote Sensing of the environmental an Earth resource perspective. Prentice Hall, Upper Saddle River, New Jersey.

Kala, C. P., Dollo, M., Farooquee, N. A. and Choudhury, D. C. (2008) Land-use management and wet-rice cultivation (*Jebi Aji*) by the Apatani people in Arunachal Pradesh, India. *Outlook on Agriculture.* 37**(2):** 125-129

Kumar, N. (2004) Primary education in India, initiative and prosoects. *Kurukshetra* **53(8):** 9-14

Santaram S. Oinam, Jagdish C. Kuniyal, Yaswant S. Rawat, S.C.R. Vishvakarma and Dinesh C. Pandey (2007). Disrupting Land-use changes cause land degradation in Fragile cold desert ecosystems: A case of the Lahaul valley in Northewester Indian Himalaya. *Journal of Water and land-use management.* 7**(2):** 197-217

Santaram S. Oinam, Yaswant S. Rawat, Khoiyangbam, R. S. Khwairakpam Gajananda, Jagdish C. Kuniyal and S.C.R. Vishvakarma (2005). Land-use and land-cover changes in Jhalma watershed of the Lahaul valley, cold desert region of the Northwester, Himalaya, India. *Journal of Mountain Science.* 2**(2):** 129-136.

Saxena, K. G., Rao, K. S., and Kothyari, B. P. (1990) Social forestry in a broad perspective of integrated resource management and sustainable production. *International Journal of Ecology amd Environment Sciences.* **16:** 15-26.

Whyte, R. O. (1961) Evaluation of land-use in south western Asia; In: A history of land-use in arid regions (Eds., L. D. Stamp) UNESCO.

7

Diversifying Land-Use: Case of Shivaliks in Indian Himalayas

Sanjay Arora & M.S. Hadda

INTRODUCTION

The foothills of Shivaliks extending from Jammu and Kashmir to Uttarakhand states cover a total area of 3.10 mha. The annual rainfall in the region is approximately 800-1250 mm of which 80 per cent occurs in monsoon season. Soil erosion through water is one of the major problems in the area during monsoons and moisture stress for winter crops (Arora, 2006). Rainfed areas face multifaceted problems of poor soil fertility, erratic rainfall, low soil moisture, poor economic condition of the farmers who cannot afford to invest adequately. This tends to diminishing livelihood support system. Hence, there is an urgent need to evolve land-use diversification systems in rainfed areas to avoid land degradation, conserve and develop natural resources and increase the living condition of the farmers (Hadda *et al.*, 2007).

Land-use diversification involves multidisciplinary approach including interaction among climate, soil, water, vegetation, livestock and socio-economic dimension in devising most remunerative, eco-friendly and environmentally sustainable land-use.

The various approaches for sustainable land-use diversification in the rainfed areas include characterization of socio-economic and biophysical resources, integrated watershed development, rainwater harvesting and improving rainwater use efficiency, contingency crop planning to minimize crop losses during the period of climatic aberrations, diversification of crop production systems by including legumes, medicinal and aromatic plants, poultry, livestock farming to minimize climatic risks (Sharma, 2004). Alternate land-use *viz.* agro-forestry, agri-horticulture, silvipastoral and agri-horti-pastoral systems to maximize returns and integrated nutrient, soil, water and crop management as well as soil productivity enhancement by incorporating organic matter. It also includes designing appropriate and cost effective farm implements for saving time during agricultural operations, initiation of crop insurance scheme in rainfed areas, improving credit availability and input supply systems, setting of agro-industry in rainfed areas and efficient remunerative intercropping and sequence cropping systems with emphasis on pulses and oilseeds for different agro-climatic regions.

Since drought is a common evil in rainfed areas, appropriate techniques and technologies are to be adopted on watershed basis for effective soil and water conservation to have sustainable land diversification systems and improve the socio-economic condition of the poor farmers.

Agriculture in Foothills

The hill and mountain areas differ from the plains in topography, elevation, physiographic feature, diversity of habitats for flora and fauna, ethnic diversity, land-use systems and socio-economic conditions. Hills, in general, offer a vast scope for cultivation of diverse mix of crops - cereals, pulses, oilseeds, vegetables, flowers and fruits. Animal husbandry is an integral part of farming system. Temperate climatic condition on high altitudes enables to grow some fruit and vegetables, which otherwise cannot be cultivated in plains and the produce of which provides much dividends to the cultivators in hills thereby raising their economic status (Gupta *et al.*, 2000). However, due to increasing population pressure, the land on the hills in general, is not being cultivated on the scientific principles. Marginal and sub-marginal lands have been brought under plough without taking adequate conservation measures. This has led to the degradation of production resources (land, water and vegetation) causing devastating floods in the plains destroying crops, buildings, enterprises and adversely affecting human being because of silting of the riverbeds, reservoirs and other water bodies (Gupta, 2005).

Agricultural development in hills is constrained by lack of appropriate technology, inadequate infrastructure, poor input-output marketing, insufficient extension education and network and decrease in area for forest dwellers and great diversity in farming systems/land-use type. Traditional agriculture is practiced which combines crop production with livestock, horticulture and farm forestry/agro-forestry in an integrated fashion, which is adapted to the limitations of climatic and edaphic conditions (Sharma, 2004).

Inaccessibilty, fragility, marginality, heterogeneity, natural instability and human adaptation mechanism are mountain specificities, the identified key factors, to be addressed for sustainable agricultural development in hills (Arora *et al.*, 2006). The major components of hilly agro-ecosystems are crop production, livestock, horticulture and farms/agro-forestry. To meet the demands of increasing population, we have to conserve the natural resource base and ensure its optimal utilization for attaining sustainability. The resource base in agricultural production system includes land, air, water, energy and upgradation of biogenetic diversity. Land and water are the most important factors in agricultural production and the production in plains too is dependent upon the management of agri-geo-biotic factors in hills/mountains. If the natural resources are managed effectively, efficiently and optimally in the hills, they will positively influence land and water resource by regulated water supplies and undisturbed soil base in the plains.

The foothills of Shivalik region consists of 3 sub-zones of Jammu and Kashmir, Himachal Pradesh and Uttarakhand Hills. The region consists of skeletal soils of cold region, podsolic soils, mountain meadow soils and hilly brown soils; lands of the region have steep slopes in undulatin terrain. Soils are generally silty loam with altitudinal variations. They are prone to erosion hazards, and slides and slips are quite common.

The forest area in Jammu and Kashmir and Himachal Pradesh is much less compared to Uttarakhand hills. The cropping intensity is lowest in Jammu and Kashmir and highest in Himachal Pradesh. The productivity of all the crops (except maize in Himachal Pradesh) is lower than the all India average and annual growth rate is much less.

The states falling in this region have been further classified into sub-agro-climate zones based on the criteria of homogeneity in characteristics such as rainfall, temperature, topography and the specific features (Table 7.1).

This region, in general, is divided into 4 sub-regions and vegetation varies from low altitude hills to high mountains.

Table 7.1: Vegetation in the sub-regions of hills

Sub-region	*Crops/Forest Type*
Low altitude sub-tropical zone of J&K, submontane and low hills sub-tropical zone of HP and sub-tropical zone of Uttarakhand	Crops: Wheat, maize, rice, sugarcane, Vegetables, pulses (gram and lentil) Horticultural crops: Citrus, mango, litchi, peach, plum, guava, ber, olive, almond, pistachio, pomegranate Forests: Deciduous forests, dry deciduous shrubs and grasses.
Mid-to-high altitude intermediate zone of J&K, mid hills and sub-humid zone of HP and sub-temperate zone of Uttarakhand	Crops: Rice, wheat, cowpea, sesame, vegetables. Horticultural crops: Stone fruits, citrus, walnut Forests: Lower west Himalayan temperate forests and Himalayan Chir pine forests Plantation crop: Tea
Mid to high altitude temperate zone of J&K, High hills temperate zone of HP and temperate zone of Uttarakhand	Crops: Maize, rice, barely, oilseeds, pulses, off-season vegetables Horticultural crops: Apple, pear, plum, kiwi, peach, apricot, cherry, chestnut.
Cold arid zone of J&K High hills temperate dry zone of HP and Alpine zone	Crops: Barely, maize, pulses, miner millets, kuth, hops, cumin, saffron, cabbage seed, sugarbeet, chicory, potato Horticultural crops: Apple, nuts, chilgoza, neoza pine Forest: alnus, ulnus, cettil. Salix (silver fir), silver birch and junipers Fodder: Alfalfa

Soil Resources

The foothill region is endowed with a wide range of physiographies/landforms, climate, vegetation and geology. The soils in general are light textured and prone to erosion. The slope of the agricultural fields ranged from 2 to 6 per cent where maize and wheat are the dominant crops. Because of the undulating topography, much of the monsoon rain is lost as runoff through a mesh of streamlets (*choes*). The dominant soils are classified under six orders *viz.* Affisols, Aridisols, Entisols, Inceptisols, Mollisols and Vertisols. Inceptisols occupy the largest area (43.8 per cent) followed by Entisols (28 per cent). Most of the soils in the area are represented by the great groups of Ustifluvents. Ustipsaments and Ustocrepts with local Haplustalfs and Fluva-quents. The soils in the region have no calcic horizon but have an ochric epipedon and below it an argillic horizon (Teli, 1995).

Land-Use Diversification

Land-use diversification involves multidisciplinary approach including interaction among climate, soil, water, vegetation, livestock and socio-economic dimension in devising most remunerative, eco-friendly and environmentally sustainable land-use.

Sustainable Land-Use Diversification

The various approaches for sustainable land-use diversification in the rainfed foothill areas include:

1. Characterization of socio-economic and biophysical resources.
2. Integrated watershed development.
3. Rainwater harvesting and improving rain water use efficiency.
4. Contingency crop planning to minimize crop losses during the period of climatic aberrations.
5. Diversification of crop production systems by including legumes, medicinal and aromatic plants, poultry, livestock farming to minimize climatic risks.
6. Alternate land-use *viz.* agroforestry, agri-horticulture, silvipastoral and agri-horti-pastoral systems to maximize returns.
7. Integrated nutrient, soil, water and crop management.
8. Soil productivity enhancement by incorporating organic matter.
9. Designing appropriate and cost effective farm implements for saving time during agricultural operations.
10. Initiation of crop insurance scheme in rainfed areas.
11. Improving credit availability and input supply systems.
12. Setting of agro-industry in rainfed areas.
13. Efficient remunerative intercropping and sequence cropping systems with emphasis on pulses and oilseeds for different agro-climatic regions.

Benefits of Diversified Land-Uses

It is often emphasized that second green revolution has to be in rainfed areas; it is hard to realize unless location specific farming system based strategies are evolved. We need to focus on grain/fodder, crop + sheep/goat rearing + agroforestry (tree – crops, agri-silvi, agri-horticulture), agri-dairying and agri-apiculture systems for the rural livestock population to avoid denudation of rainfed foothill areas.

Putting rainfed foothill areas under horticultural and other alternate high value crop, yields high dividends. Such crops which can be grown in these areas are ber, aonla, guava, mango, pomegranate, jamun, phalsa, sapota etc.

Some of the farmers in the foothills of Punjab have started using their land for fruit crops like aonla, mango, kinnow and guava. There is need to convince the farmers through demonstrations or 'Kisan melas' for adopting the diversified use of their land in the area to avoid risk.

Different Agri-Horticultural Systems

Cultivation of medicinal and aromatic plants is to be given high priority. These are high value dollar earning crops with less care. The crops which can be grown under dryland conditions include senna, ashwagandha, aloe, palmarosa, lemon grass, basil, indigo, henna etc. These crops can be grown profitably for the reasons that neither they are stolen nor damaged by animals, have longer shelf life, low incidence of pests and diseases, they are export oriented, generate employment through out the year, low labour requirement and many of these crops can be included in intercropping systems with field crops.

Livestock rearing is an important source of income and means of employment for a large number of farmers in rural India and in rainfed foothill region of Punjab. About 23 per cent of households owing milch animals are landless and some 30 per cent are marginal farmers with landholding less than 1 ha. Intensification of crop-livestock mixed systems and livestock especially ruminants will continue to play vital role. It will increase both short term benefits and long term sustainability through optimum utilization of available resources.

Since drought is a common evil in the drought prone foothills, appropriate techniques and technologies are to be adopted on watershed basis foe effective soil and water conservation to have sustainable land diversification systems and improve the socio-economic condition of the poor farmers.

Agroforestry: Alternate Land-use System

Alternate land-use system may be defined as a perennial system or practice which is adopted to replace or modify the traditional land-use. Its main aim and objective is to match the land capability classes for generating more assured income with minimum risk through efficient utilization of available resources. Such systems have more potentiality and flexibility than the traditional crop production systems.

In the light of various needs of the peasants such as food, fodder, fuel, wood, fibre and fertilizer, etc., tree farming, growing of grasses, production of horticultural/

olericultural crops besides agricultural production, are different versions of need based alternate land-use system. Growing of agricultural crops alongwith trees in the same field is called "agroforestry". There are a number of agroforestry practices, like alley cropping, silvipastoral system, agri-horticulture system, horti-silviculture system, etc. The advantages of agroforestry and details of various agroforestry system have been mentioned hereunder:

Agroforestry and its Advantages

Agroforestry includes land management technique, which implies the combination of both *i.e.*, growing of crops, fruits, forest tree species and grasses. So, it is called agro-silviculture (growing of crops and forest trees), agro-horticulture (growing of crops and fruit trees), agro-pastoral (growing of crops and grasses).

Agroforestry is also called mixed cropping. In developing country like India, agroforestry or mixed cropping system is a complex and baffling task. Its complexity lies on small and fragmented landholdings, limited resources and endless hardships of diversified nature of rural families. Agroforestry system is baffling or difficult because the rural people are traditional minded, orthodox and resist any kind of change. Despite these facts, this system of farming is being popularised considerably due to its number of advantages. Some of them are as follows:-

1. The agroforestry system being the most sustainable alternative protects the natural resources and environment, which are the need of the hour.
2. This system of farming assists in checking soil erosion and landslides/ landslips.
3. Provides the vital needs of the farmers in respect of firewood, timber and the related non-wood forest products particularly in arid and semi-arid regions.
4. Adoption of agroforestry systems promise employment to the rural youth in the activities of animal production, collection, processing, manufacturing value added products from the trees and grasses.
5. Trees have a system of recycling of nutrients. Though major recycling is done through decomposition of leaves and decay of roots, yet deep-rooted plants derive the nutrients from the lower depths of the soils, making the surface soils more rich in them.
6. Leguminous trees, and shrubs and trees also enrich the soil through fixation of nitrogen from the atmosphere.
7. Covers the risks of the farmers during drought. If one crop fails due to lack of rains, the other succeeds.

Land Shaping and Diversion of Water

An area having 6 percent or more slopes should be put under arable crop production only after proper land shaping with provision for disposal of clean up land water plus one coming from area under cultivation. Thus on areas up to even 10 per cent slope contour bunding might serve the purpose but beyond this limit bench terracing has to be adopted. Soil depth is an important factor when undertaking construction of terraces up to 50 per cent slope. Inward terraces be constructed and conditions created so that as much rain water as possible can be absorbed by the soil. The terraces on the high edges should be stabilized by grasses and if possible all the area beyond 30 per cent slope be put under permanent plant cover which should consist of quick growing trees having timber or fuel value and/or bushes with tremendous canopy characterization. Lands from 30-49 per cent slope can be treated to make terraces so designed as to conserve adequate moisture to support on terrace tree growth and yield clean water without sediment load to be channeled downward for use in cultivation of crops.

Cropping Pattern

Cropping pattern on terraces has to be according to dictates of rainfall pattern, slope and length of the slope with provision that soil and required water conservation measures have been taken to fore-stall rill and gully erosion. The first strip should be of erosion resistant material i.e either a grass of fast growing legumes to be sown across the slope. Second strip be of a grain crop but sowing of this strip be also done across the slope. From 10 per cent slope, terraces are constructed to cover a slope of 50 per cent. Below 10 per cent even contour bunding can suffice provided bunds are adequately stabilized by grasses which have deep anchorage *vetiver* grass is one such material which can provide considerable stability to bunds and by providing shade it can also prevent cracking of bunds.

Land-Use Capability

Land should be utilized according to its capability class and there is every likelihood that in all watersheds there would be some lands considered as waste and degraded. These should be used to promote social forestry programme and it is now regarded as one of the forth right steps in meeting fuel, fodder, timber and other requirements.

Establishment of Grasses

Soil erosion has reached crisis proportion in the J&K State and over 80 per cent of the geographical area is affected by soil erosion in one form or the other and nearly 15 per cent of the affected area has reached a stage of complete destruction. On an average 10-30 tones of valuable top soil per hectare are lost from the fields and may be even higher in severe cases. It is estimated that about 85000 tonnes of nutrients are lost annually. Further, erosion due to land slides is a gigantic problem and one of the glaring example is of 'Nashri slip'. Our lakes, reservoirs and flood absorption basins are getting silted up at an alarming rate and that soil erosion in the catchment areas is one of the main causes of this mountainous state.

Grass cover and vegetative barriers has proved very effective in protecting the land from soil erosion. Misuse of grasslands and catchment areas as a result of continued grazing pressure has been one important single contributory factor.

Besides this grasses help in restoration of soil fertility and productivity of eroded and denuded lands and protection of soil conservation structures like bunds, terraces and water ways

Multipurpose Tree Species

The most important advantage of nitrogen fixing trees (NFT) is that many NFT species combine more than one use or product in a single tree. These products often include fuel, fodder, food, fertilizer and fibre. NFT's often have high protein foliage which can be used for fodder, green manure or human food in addition to wood which is generally suitable for fuels, timber or other uses. NFT's have been successfully used as shade or companion crop in a variety of systems for decades.

Multipurpose NFT's are multipurpose when managed for multiple outputs or services. An obvious example is found in NFT species *Leucaena leucocephala* and can easily be managed to provide a number of products and services from the same tree.

The greatest value of these trees as opposed to non NFT's, is obtained when a particular NFT can provide economically viable quantities of desired products and services and at the same time fix enough N for its own use and other purposes.

In general, the environmentally beneficial characteristics of trees independent of whether tree fixes nitrogen include

- Ability to hold soil in place
- Long term plant cover
- Ability to ameliorate the climate
- Potential for soil improvement
- Adaptability to soil and climatic conditions
- Resistance to pests and diseases
- Suitability as fodder
- Ability to tolerate common lopping, pruning and coppice management practices
- Accelerated growth rate
- Soil conservation abilities
- Multiple uses for wood
- Appropriate rooting characteristics

Tree Planting and Afforestation

Forest cover plays significant role not only in checking the soil erosion and run-off losses but also maintains the balance in physical landscape through the regeneration of degraded and eroded lands. It has also been found that natural fallows under grasses and shrubs or grasses alone have negligible run-off and soil loss than cultivated fallows which give maximum rate of run-off and soil loss. Vegetative cover increases the water stable aggregates in soil, protects it against rain drop, reduces water velocity, increases infiltration capacity and fertility of the soil, and also provides valuable food, feed/fodder for the livestock and the human population. Trees/shrubs intercept direct rainfall and reduce its velocity and force with their canopy.

Trees for Soil and Water Conservation

Soil erosion by water is one of the serious problems denuding the environment of the foothill region of Shivaliks (Arora, 2006; Hadda *et al.*, 2007). One, two or three lines of woody perennials can be planted closely along the contours and with time these hedges would develop into micro-terraces through accumulation of soil over them. They will serve as barriers to surface run-off and improving thereby, soil fertility through incorporation of their pruning. Suitable tree species are: *Leucaena leucocephala, Gliricidia* spp., *Sesbania* spp. Studies conducted in Lithosols soil of *Kandi* belt of Jammu, showed that run-off was considerably reduced due to presence of a strip of tree species across the slope (2 per cent). The run-off loss of rain water was least with *Acacia nilotica* and *Albizzia lebbeck*, while *Acacia catechu* showed moderate run-off loss but was less than the check plot, *i.e.*, without trees.

Tree Plantation

A large tract of North West Himalayan region which once was under dense forests, has now degraded under heavy biotic pressure. As a result, much of its area has been reduced to deforested terrain and scrubs. Many of the so-called deforested belts and eroded regions are prone to further soil erosion and denudation. Soil and water conservation in these areas, demand right selection of various tree species being planted under the Social Forestry and afforestation projects. Monoculture and commercial plantations should be totally stopped in these areas. The quick growing species belonging to leguminosae like *Leucaena, Albizzia, Dalbergia* and *Sesbania* are suitable for subtropical areas especially in the Shivalik hills of Jammu, Punjab and Himachal Pradesh. Black locust (*Robinnia* spp.), honey locust (*Gladitsia* spp.) as well as alders (*Alnus* spp.) grow abundantly in sub-temperate/temperate zones of North West Himalayas.

Some of the species which play an important role in providing leaf fodder besides controlling soil erosion are *Grewia optiva, Morus alba, Aegle marmelos, Bauhinia variegata*. The fuel wood species suitable for degraded and denuded areas are: *Acacia, Albizzia* and *Prosopis*. Planting of *Dalbergia sisso, Acacia catechu, Dendrocalamus* spp. and *Pinus roxburghii* in Nagrota area of Jammu Shivalik hills has shown a survival rate of above 90 per cent. *Leucaena lecocephala* and *Dendrocalamus* spp. are found useful in rehabilitation of rocky areas. Briefly the quick growing trees/shrubs for the rehabilitation of eroded/degraded areas of the North West Himalayas, are presented in Table 7.2.

Planting of Grasses and Shrubs

The grasses on their decay, add carbohydrates (simple sugars, starch, cellulose), protein and fats to the soils which promote soil aggregation. This also results in reducing soil erosion and run-off losses. Grass roots, penetrating into the soil, have been observed in two years old grasses like *Cenchrus ciliaris, C. setigerus, Lasiurus sindicus* and *Pandicum anidotale* on sandy soil in Kangra area of Himachal Pradesh. The roots of these grasses penetrated to 150, 125, 320 and 210 cm, respectively and produced lot of root matter.

Table 7.2: Vegetation suitable for the North West Himalayan region

Region	*Areas Vulnerable to Denudation/Soil Erosion*	*Type of Erosion*	*Tree Species Required to be Planted*
A. Jammu & Kashmir			
(a) Lower to Middle Himalayas	Hansgala behind Salal project, Panj Gali and Makhi Dhar comprising a number of villages of Reasi and Arnas tehsils of Udhampur district.	Apart from sheet rill and gully erosion, a number of recent slides have been taken place.	Quick growing legume species. *Acacia* spp. for the lower and *Pseudoacaci* spp. for middle Himalayas.
(b) Middle Himalayas	Bhadarwah, Doda and Kishtwar tehsils of Doda district in Pir Panjal range, Basholi, Bhadu, Billawar of Kathua as well as Poonch area.	Landslides very common, roads sometimes remain blocked upto 20 days, kill gully and ravines very common.	Encourage plantation on both sides of the roads particularly leguminous plants and other fast growing species.
(c) Trans-Himalayas/ Great Himalayas	Ladakh and parts of Doda district.	Linear erosion, landslides, nivo-glacial movement, solifluction etc.	*Salix, Junipers, Artemisia* with *Lonicera* species should be encouraged.
(d) Lower Himalayas or Shivalik Himalayas including *Kandi* belt of Jammu	Parts of Jammu, Kathua and Udhampur districts as well as of Rajouri and Poonch districts.	Sheet, rill and gully erosions as well as ravines.	Fuel wood, leaf fodder and fast growing species are preferred like *Leucaena leucocephala, Acacia* spp.
B. Himachal Pradesh			
(a) Middle Himalayas	Dhaula Dhar range covering parts of Kangra, Chamba, Kulu, Shimla and Sirmour districts.	Linear erosion, landslides, rockfall slumping, etc.	Local and quick growing species like *Populus, Eucalyptus*, etc. *Quercus* species should also be grown.
(b) Great Himalayas	Lahaul Spiti and Kinnaur.	Laminar, rill and gully erosion. Solifluction pinnacle erosion etc.	*Salix, Populus* and other local creeping and matting types.
(c) Lower or Shivalik Himalayas	The districts of Hamirpur, Una, Bilaspur and parts of Sirmour, Kangra and Solan.	Sheet, rill and gully erosion.	Local species like *Acacia, Dalbergia sissoo, Albizzia* as well as quick growing species like *Leucaena leucocephala*.
C. Uttarakhand			
(a) Middle Himalayas	Mussoorie hills, Nainital areas, Rishikesh-Joshimath road in Garhwal, Tanakpur-Tawaghat in Kumaon Pithoragarh district.	Landslides/land slips, gullies ravines, glacial erosion.	Local species should be given preference. They should be grown on either side of the road.
(b) Shivalik Himalayas	Dehradun, Mussoorie hills, Terai belt.	Sheet, rill and gully erosion.	Quick growing tree species and bushes.

Besides providing fodder, the grasses protect soil conservation structures like terraces bunds and waterways. Encouraging results of Bhabar grass (*Eulaliopsis binata*) for conserving degraded bouldry lands and stabilization of steep slopes and gullies near Dehradun. Raising jointly, grass species like *Chrysopogon fulvus* (at 0.75 × 0.75 m) and species like *Dalbergia sisso* (at 9.14 × 9.14 m) in V and VI land capability classes, have shown not only a high yield of air dry grass (10.55 t ha^{-1} $year^{-1}$) but also about 64 t fuel wood was available after 19 years. There was sufficient control of soil and run-off losses. Dhruva Narayana (1982) reported that the replacement of annual grasses such as *Aristida furiculata* and *Themeda triandra* by *Apluda mutica, Heteropogon contortus, Dicanthium annulatum* and *Cenchrus* spp. in ravine lands, has resulted in a natural reduction in run-off and soil loss alongwith improvement in the quality and quantity of grass yields.

In the degraded areas of Garhwal Himalayas, planting of some shrubs like *Eupatorium triplenerve, Woodforadia fruticosa, Desomodium* spp., *Boehmeria* spp., *Berberis* spp. and *Rumex hastatus*, have been found suitable in controlling soil and water conservation.

In a bid to stop soil erosion in eroded areas and community wastelands of Himachal Pradesh, the farmers have been encouraged to grow shrubs such as *Mimosa himalayana, Desmodium* spp., *Carissa spinarum* and *Adhatoda vasica* for the Shivalik Himalayas and various *Quercus* species, and coniferous for middle Himalayas (Sharma *et al.*, 1998). These shrubs serve the dual purpose of meeting the villagers firewood needs and helping soil and water conservation by preventing run-off. Besides, efforts were made to grow grass species such as *Cenchrus* and *Stylosanthes* to supplement the fodder and strengthen the soil.

References

Arora, Sanjay, 2006 "Preliminary assessment of soil and water conservation status in drought prone foothill region of north-west India". *Journal of World Association of Soil Water Conservation* J1-5, pp. 55-63.

Arora, Sanjay, Sharma, V., Kohli, A. and Jalali, V.K., 2006. "Soil and water conservation for sustainable crop production in *Kandi* region of Jammu". *Journal of Soil and Water Conservation, India* **5(2)**: 77-82.

Dhruva Narayana, V.V., 1982. "Souvenir. 12th International Congress Soil Science", Indian Society Soil Science, 8-16 Feb., New Delhi.

Gupta, R.D., 2005. "Soil erosion: Causes, hazards and control measures", in *Environmental Degradation of Jammu and Kashmir Himalayas and Their Control*, Associated Publishing Company, Karol Bagh, New Delhi.

Gupta, R.D., Sharma, P.D. and Verma, K.S,.2000. "Environment and ecology of cold arid region – A case study of Lahaul and Spiti in Himachal Pradesh", in J.P. Sharma and A. Aziz Mir (Eds.), *Dynamics of Cold Arid Agriculture*, Kalyani Publishers, Ludhiana.

Hadda, M.S., Arora, Sanjay and Bhardwaj, D.D., 2007. "Assessment of Erosion Risk of a Catchment Using Soil and Rainfall Information in Foothills of Punjab". *Indian Journal of Dryland Agricultural Research and Development*, **22 (2)**: 70-74.

Sharma, P.D., 2004. "Managing natural resources in Indian Himalayas". *Journal Indian Society Soil Science*, **52(4)**: 314-331.

Sharma, P.D., S.P. Sharma, K.D. Sharma, and R.S. Minhas, 1998. "Status report on catchments of Giri and Batq rivers in Himachal Pradesh". Department of Soil Science, H.P.K.V.V., Palampur.

Teli, B.L., 1995. "Soil resources: Problems and planning in hills", in B.L. Koul (Ed.), *Ecodegradation of Himalayas – Causes and Remedies*, Vinod Publishers and Distributors, Pacca Danga, Jammu, pp. 121-129.

8

Environmental Management Planning: Remote Sensing and GIS Application

D.K. Agrawal & M.S. Lodhi

INTRODUCTION

Environmental Impact Assessment (EIA) is a decision making tool to ensure consideration of environmental impacts of a developmental activity through incorporation of appropriate Environmental Management Plans (EMP). In India, EIA was introduced in 1994 with Ministry of Environment and Forest (MoEF) introduced EIA notification under Environmental Protection Act (EPA), 1986 and made EIA mandatory for 29 highly polluting activities. The effectiveness of any EIA document depends on the accuracy/certainty of existing environmental baseline which is considered as ground for modeling the future environmental scenario because in turn the effectiveness of EMP would rely on these predicted impacts only. The environmental baseline studies include primary data sets on each and every environmental parameter which is a time taking and cumbersome process. The analysis and handling of large volume of such data sets with time-space is further a massive task to do, thus there is an earnest need of a tool or support system which can handle large volume of data sets and explicitly analyze the same. Remote Sensing (RS) and Geographical Information System (GIS) are such recently developed tools, loaded with all the above facilities in addition to its biggest asset of presenting the large databases in an overlay format for better understanding the interactions among the datasets of various parameters. RS techniques can play an important role in generating baseline information on soils, slope, forest cover, surface water, groundwater and land-use/land-cover and subsequent integration of such information with socio-economic database in GIS platform.

At present, various remote-sensing satellites are capable of providing extraordinary data on all aspects of the environment with microscopic resolution. These inestimable spatial data sets can be utilized in environmental decision making processes. GIS is a computerized resource database system, which is referenced to geographical coordinate system. It is a tool to store, select, explore, analyze, and display various spatial data sets. Utilizing these abilities of GIS, the tedious process of baseline data storage and subsequent analysis can be made efficient and less time consuming. In recent years, the advent of user friendly geographical information systems (GIS) and availability of remote sensing (RS) data has expanded scope for application of these techniques in EIA studies which has further augmented the capabilities for better formulation of EMP. During last two decades, several researchers have recognized and advocated for the use of Geographical Information System (GIS) in all stages of EIA studies and EMP formulation as a valuable tool (Goodchild, 1993; Nyerges, 1993; Eedy, 1995; Antunes *et al.*, 1996; Berry, 1996; Joao and Fonseca 1996; Treweek and Veitch 1996; Haklay *et al.*, 1998), however very little evidence is available on actual use of these techniques except for developing base maps, etc. The appropriate use of such approach can undoubtedly improve the qualitative substance of the EIA document and result in better EMP formulation. The present paper provides practical aspects of use of RS-GIS based approaches in EIA and EMP formulation in respect of a proposed hydropower project in Uttarakhand state of India to highlight the effectiveness of the tools in formulating EMP that are more linked to primary data sets on various environmental attributes.

Case Study: Vyasi Hydropower Project

The project under study is proposed to be situated in Garhwal Himalayas of Uttarakhand state, which is highly sensitive in terms of various attributes of environment. The Vyasi dam site (30°31′N, 77°54′E) is 66.6 km from the state capital Dehradun and is approachable through passing National Highway 123. The project envisages construction of 86 m high concrete gravity dam and appurtenant structures to generate 120 MW of electricity (DPR, 2007). In addition to dam, project includes construction of 2.7 km long tunnel (7m circular diameter), a surface power house as major structures. The construction of Vyasi project will create a reservoir upstream of the proposed dam, consequently an area of 96.729 ha including riverbed (29.306 ha) will be submerged. The project area has divergent topography with steep slopes. The Vyasi reservoir would have a surface area of 70.3 ha and will store 12.6Mcum of water upto Full Reservoir Level (FRL) at an elevation of 631.5m. Overall 135.425 ha of land is required for the construction of the project. Figure 8.1 presents the location map of the Vyasi dam and the entire Yamuna catchment upto the proposed dam site. The primary socio-economic survey (GBPIHED, 2007) in the submergence area revealed that a population of 749 peoples (137 families) will be affected as per the 'definition of affected family' given by Ministry of Rural Development (Govt. of India).

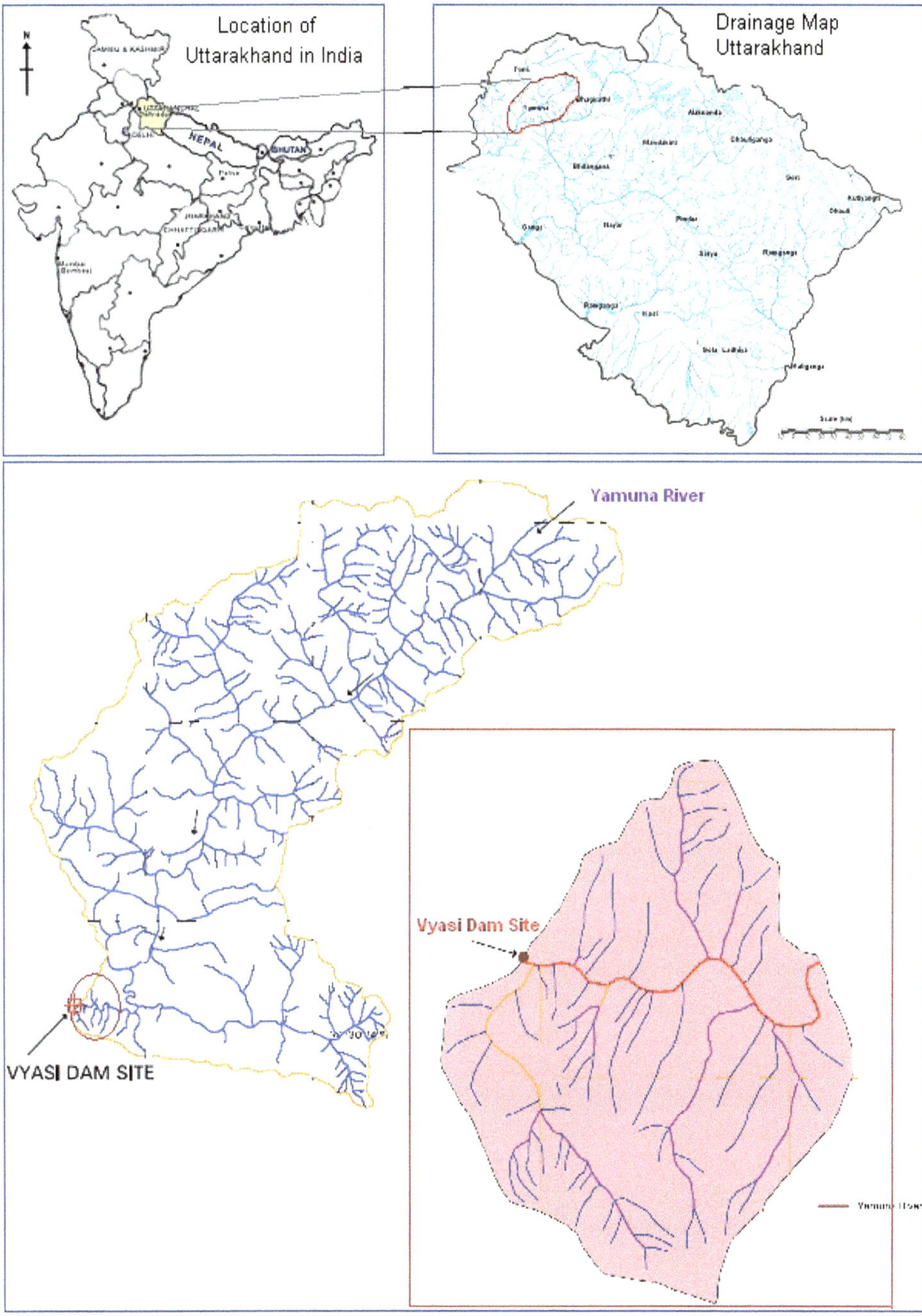

Figure 8.1: Location map of Vyasi Project and its free draining catchment, influence zone and submergence zone.

EIA study being a multidisciplinary process by nature requires vast as well as reliable database on several unique environmental aspects which are of particular importance. The procurement of baseline database using GIS can improve the reliability and accuracy of the results apart from reducing effective time incurred in data collection as compared to other methods. In further discussion, the methods and procedural explanation is presented for baseline data generation for EIA study in the context of proposed hydropower project.

Delineation of Free Draining Catchment; Influence Zone and Submergence Zone

In order to conduct EIA/EMP study for Vyasi HEP, the project area was delineated into:

Free Draining Catchment Area of Vyasi Dam

There are two consecutive proposed hydroelectric schemes Lakhwar and Vyasi, to be constructed on Yamuna River by NHPC. Here it is pertinent to point out that as per the standard definition of free draining catchment area, for practical purposes of developing EMP the catchment area between proposed Vyasi and Lakhwar dam sites is considered as the free draining catchment for Vyasi project. Adopting hierarchical delineation system developed by AISLUS (AIS and LUS Technical Bulletin - 9), Vyasi HEP falls in the Upper Yamuna Catchment from confluence with Hindon (Code 2C6) with basin code as 2C for Yamuna and region code as 2 for Ganges. The sub-catchment code is 2C6C for uppermost Yamuna in Himalayan Mountains, whereas the watershed code is 2C6C2 for main Yamuna River. Specifically, the free draining catchment area of the Vyasi HE project falls under the sub-watershed (2C6C2f) of watershed 2C6C2. Looking at the drainage network between these two dam sites, it is seen that there is no major tributary joining the river Yamuna and therefore as such the free draining catchment area is 1361.37 ha only (Figure 8.1).

Influence and Submergence Zone of Vyasi Project

The zone surrounding to the project site is particularly expected to be environmentally influenced by the proposed activity and is therefore considered as the influence or impact zone. As per the EIA guidelines, an area of 7 km radius from the project site is considered as influence zone for a particular project, but here in the present study a different approach was followed for influence zone demarcation. With the help of GIS, using DEM and digitized drainage network of the catchment area, the boundary of the likely reservoir was delineated in upstream of the dam considering Full Reservoir Level (FRL) elevation as the submergence levels in the free draining catchment (Figure 8.1 for submergence zone boundary). Subsequently, the influence zone was generated by extending the reservoir boundary in all directions with 7 km perpendicular aerial distance (Figure 8.1). The influence zone so demarcated has an area of 16, 943.39 ha.

Data Used and Methodology

Various base maps were collected from different secondary sources *viz.* Survey of India Topo-sheets, National Bureau of Soil Survey and Land-use Planning (NBSS and LUP) Soil map, Geological Survey of India, Forest Department, Irrigation Department, etc. The printed/drawing maps were than scanned, digitized and registered as per requirement and different GIS theme layers were developed. Point data sets on roads and settlements were also collected and overlaid in the RS-GIS database.

Land-Use/Land-Cover Map

Remotely sensed satellite images (LANDSAT) were acquired for the year 2002 (Row-39, Path-146, and Date of Pass- 25 Nov 2002). Nearest neighbour re-sampling technique was used for land-use application, since the image is to be classified later. This technique simply chooses the actual pixel that has its centre nearest to the point located in the image. The pixel is then transferred to the corresponding grid location on the topo-sheet. The original brightness value of the pixel is thus maintained. Prior to ground truthing, the satellite data was classified using supervised classification technique. The various forest types and land-cover categories as delineated for the free draining catchment is given in Figure 8.2. The various forest types and land-cover categories delineated in the map comprise of agriculture, forests and their

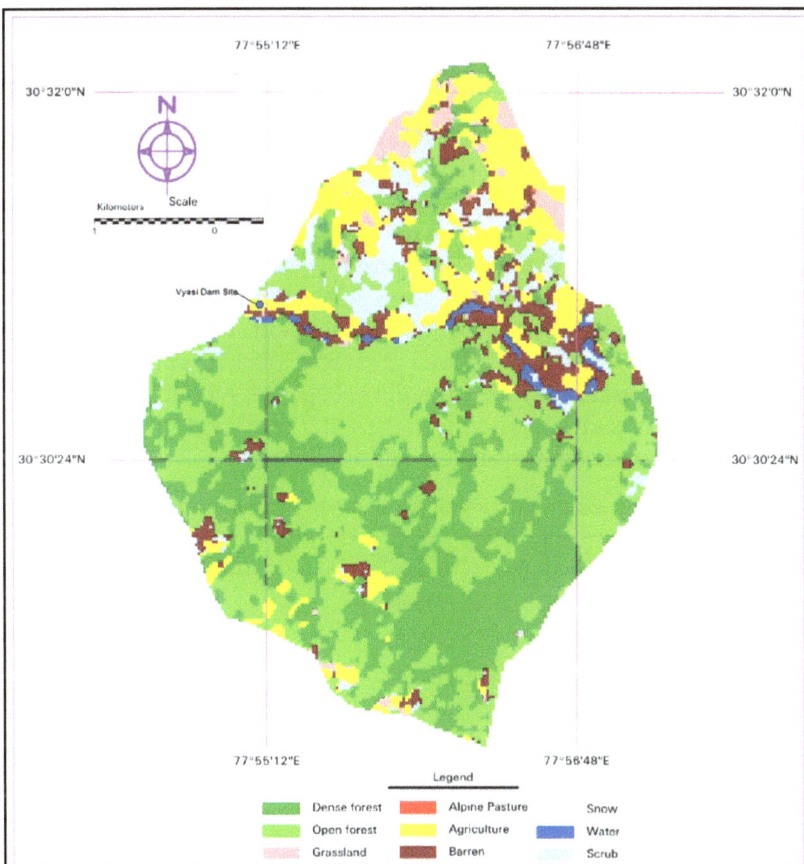

Figure 8.2: Land-use/land-cover map of the free draining catchment area of Vyasi HEP.

types and density, pasture/grazing land, and barren land. Among the different land-use/land-cover classes, maximum area is under forest. Among the various forest types, sal-mixed forest has maximum presence followed by Pine, Oak-pine mixed and Temperate broad leaved (Table 8.1).

Table 8.1: Land-use/land-cover statistics of the free draining catchment area of Vyasi HEP

Sl.No.	*Land-Use Type*	*Area (ha)*	*Percentage Area (per cent)*
1.	Dense forest	437.96	32.17
2.	Open forest	622.37	45.72
3.	Grassland	30.49	2.24
4.	Agriculture	161.55	11.87
5.	Barren	61.66	4.53
6.	Water	1.15	0.08
7.	Scrub	46.19	3.39
	Total	1361.37	100.00

Slope Distribution Map

The slope plays a great role in the loss of soil and water from an area and influences its land-use capability. Together with the nature and texture of soil, it also determines the erodibility of the soil. Slope map for the free draining catchment area was generated through standard Triangulation Irregular Network (TIN) techniques using digitized contours at 20 m interval derived from Survey of India 1:50,000 scale topographic sheets. The raster slope map was converted to vector form for further use. A perspective view of developed Digital Elevation Model (DEM) of Vyasi free draining catchment is presented in Figure 8.3. As seen from the Table 8.2, 68.06 per cent of the area has a slope of more than 50 per cent, which reflects towards the ruggedness of the area. The slope plays a great role in the loss of soil and water from an area and influences its land-use capability. Together with the nature and texture of soil, it also determines the erodibility of the soil. There is negligible zone that can be categorized as level or nearly level (very gently sloping).

Table 8.2: Slope categories in free draining catchment of Vyasi HEP

Sl.No.	*Slope Class*	*Area (ha)*	*Percentage (per cent)*
1	Gently sloping (up to 15%)	118.68	8.72
2	Moderately sloping (15%–30%)	125.61	9.22
3	Moderately steep (30%–50%)	190.61	14.00
4	Strongly sloping (>50%)	926.47	68.06
	Total	1361.37	100.00

Soils Map

Soil map (Figure 8.4) was digitized and produced using soil map collected from National Bureau of Soil Survey and Land-use Planning. The Soil map of Uttaranchal

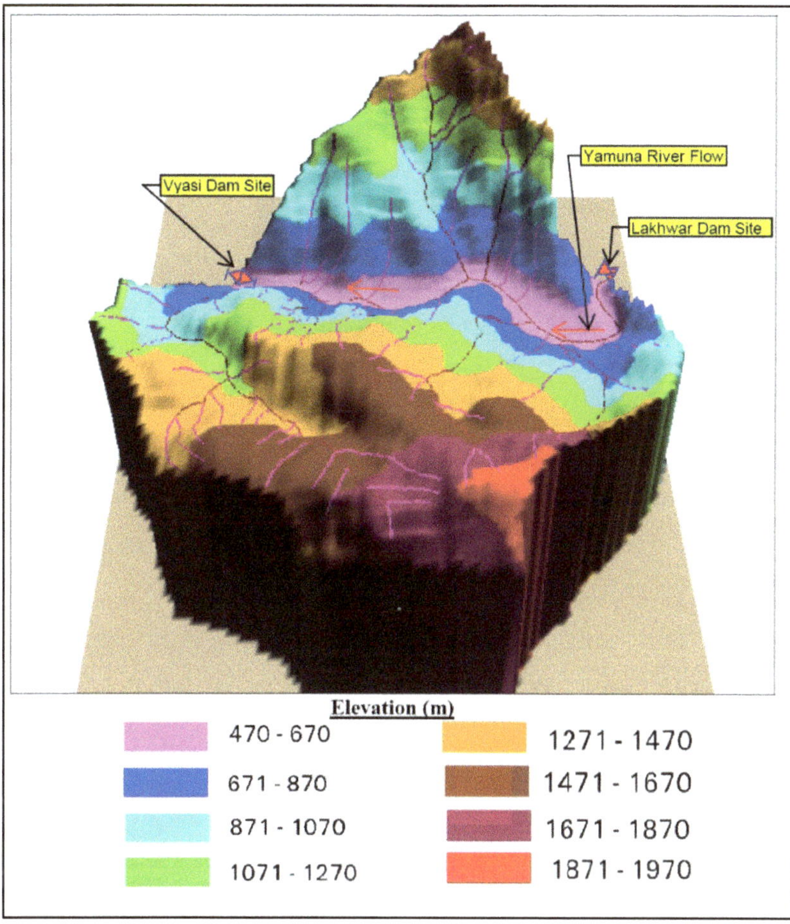

Figure 8.3: DEM of free draining catchment (false colour composite draped) of Vyasi HEP.

prepared by NBSS and LUP (Headquarters at Nagpur) was used for superimposing of results. From pedological point of view the piedmonts are the most common geomorphic unit seen in this portion of the Vyasi catchment, mostly in the middle slopes (NBSS and LUP, 2002). Gradual erosion of surface layers from slopes and its subsequent deposition in gullies has led to the formation of skeletal soils embedded with stones and boulders. Where the slope is highly steep, there is no soil - only the bedrock is starkly exposed. Steep slopes and exposed ridges house lithosols, as the layer of weathered rock is very thin. It takes the lichens and mosses to hold these lithosols. The soil type of the catchment is moderately shallow, stony and dry.

Land-Use/Land-Cover Changes: Analysis of Forest Cover Change

In the present study, the satellite images were analyzed over a period of 12 year (1991-2002) for the free draining catchment area of Vyasi HEP. RS-GIS maps were developed at 1:50,000 and analyzed for land-use and vegetation change over a period of 12 years (Table 8.3). Major negative changes are seen in dense forest, which has

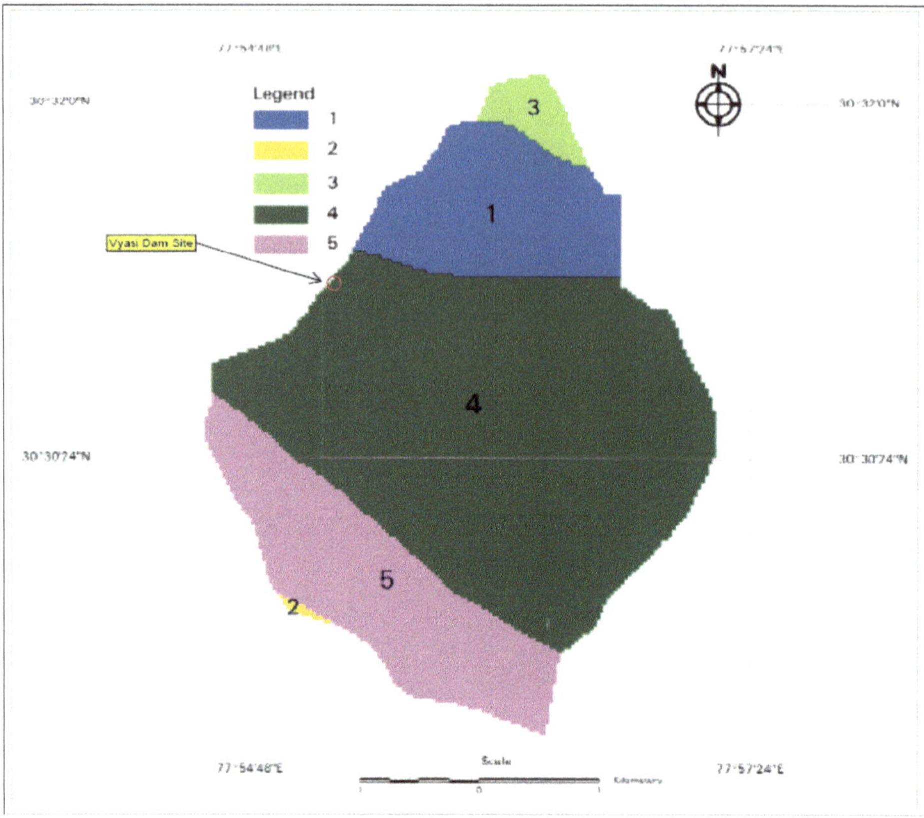

Soil Code	Soil Classification	Soil Depth	Erosion Tendency	Area (ha)
1.	Loamy Soil (Typic Udorthents)	Moderately Shallow	Moderate	204.56
2.	Loamy Skeletal Soils (Typic Udorthents)	Moderately Deep	Moderate	2.31
3.	Coarse Loamy Soil (Typic Udorthents)	Moderately Shallow	Moderate	34.54
4.	Loamy Skeletal Calcareous Soil (Dystric Eutrochrepts)	Moderately Shallow	Moderate	839.92
5.	Loamy Soil (Lithic Udorthents)	Shallow	Sever	280.04
Total				1361.37

Figure 8.4: Soil map of the free draining catchment area of Vyasi HEP.

now turned into open forest (15.54 per cent) indicating a substantial loss of forest covers; whereas it must be mentioned that probably the efforts of the Forest Department have also helped in converting almost 4 per cent of open forest in to dense forest apart from bringing a substantial 9.85 per cent scrub land in to the open forest category. Undoubtedly, such kind of analysis plays an important role in framing appropriate management strategies for various land-use classes.

Use of RS-GIS Data for Baseline EIA Study

Biodiversity Assessment

After the reconnaissance field survey and looking at the vast stretch of the influence zone of the project it was felt that it is extremely difficult to conduct primary

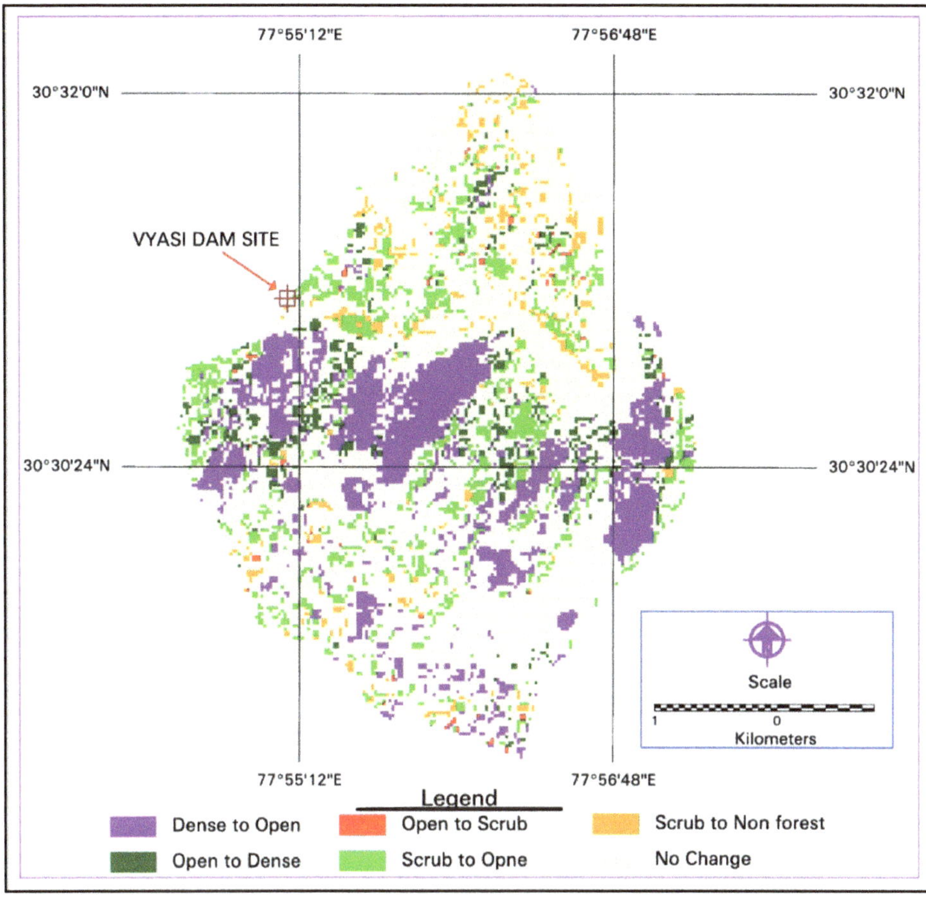

Figure 8.5: Land-use/land-cover changes over a period of 12 years (1991-2002) in the free draining catchment of Vyasi HEP.

Table 8.3: Land-use/land-cover changes over a period of 12 years (1991-2002) in the free draining catchment of Vyasi HEP

Change	*Area (ha)*	*Per cent Change in Area (Base 1991)*
Dense forest to open forest	211.5072	15.54
Open forest to dense forest	53.9136	3.96
Open forest to scrub land	4.8384	0.36
Scrub land to open forest	134.0928	9.85
Scrub land to non forest uses	48.8448	3.59
No change	908.173	66.71
Total	1361.37	100

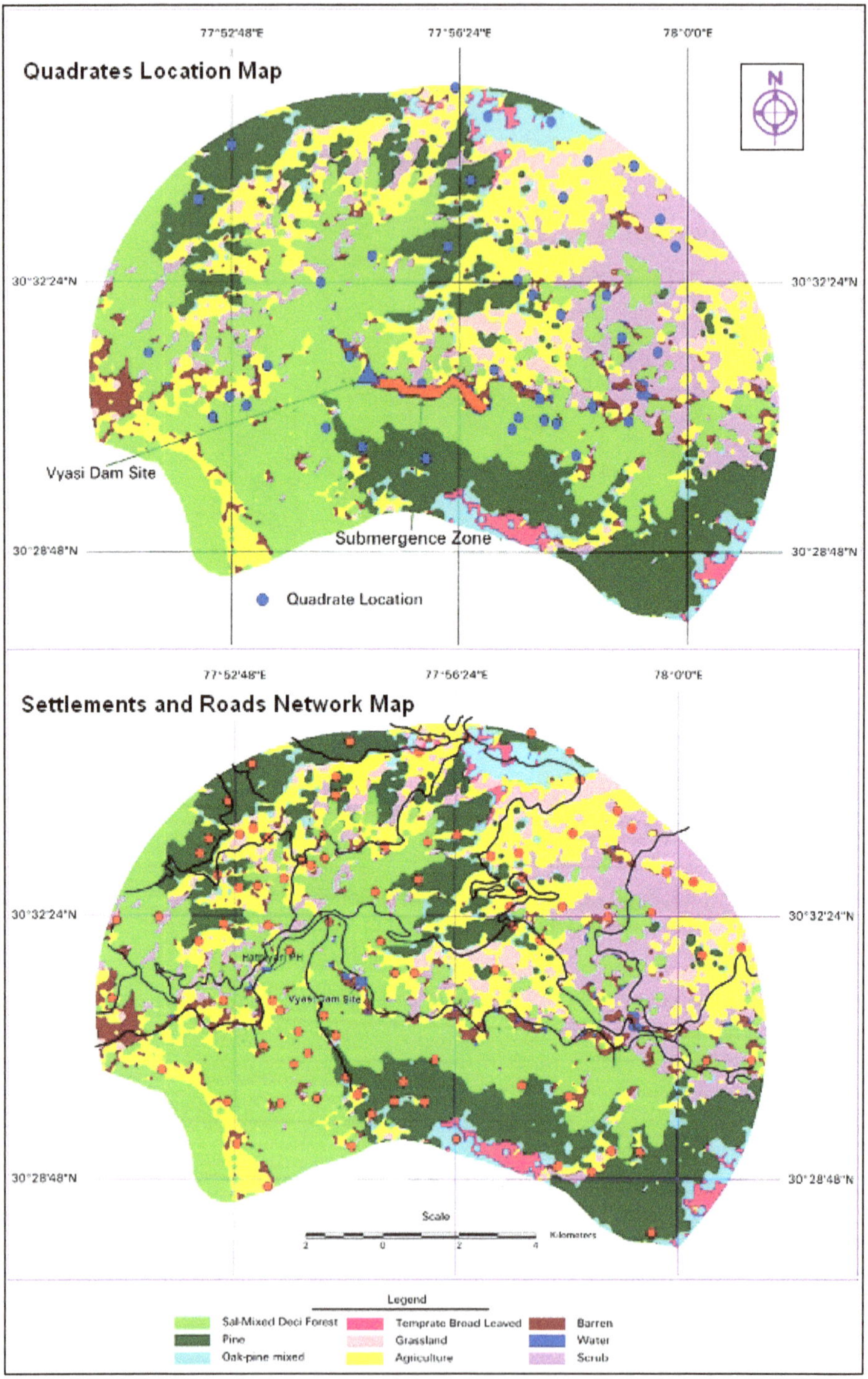

Figure 8.6: Land-use/land-cover map and location map of quadrates used for biodiversity assessment of the influence zone of Vyasi HEP along with roads network and settlements.

biodiversity assessment based on sufficient number of sampling locations (quadrates) following the general procedures. Subsequently, stratified random sampling method for biodiversity assessment was found to be best suited for the study area (influence zone). Here, RS-GIS approach played a key role in selecting the appropriate locations of laying quadrates such that they provided efficient coverage of all forest and land-use covers in all slope and aspect categories. For this purpose, the slope, aspect, forest types and other land-cover layers were overlapped and depending upon the area weights of individual land-use/land-cover class, proportionate number of quadrates were worked out. Subsequently, the calculated number of quadrates randomly marked on the land-use/land-cover map, and the composed map is presented as (Figure 8.6) and subsequently this information was used during the primary quadrate location with the help of Geographical Positioning Systems (GPS) devices. A total of 46 quadrates were thus laid in the influence zone and submergence zone.

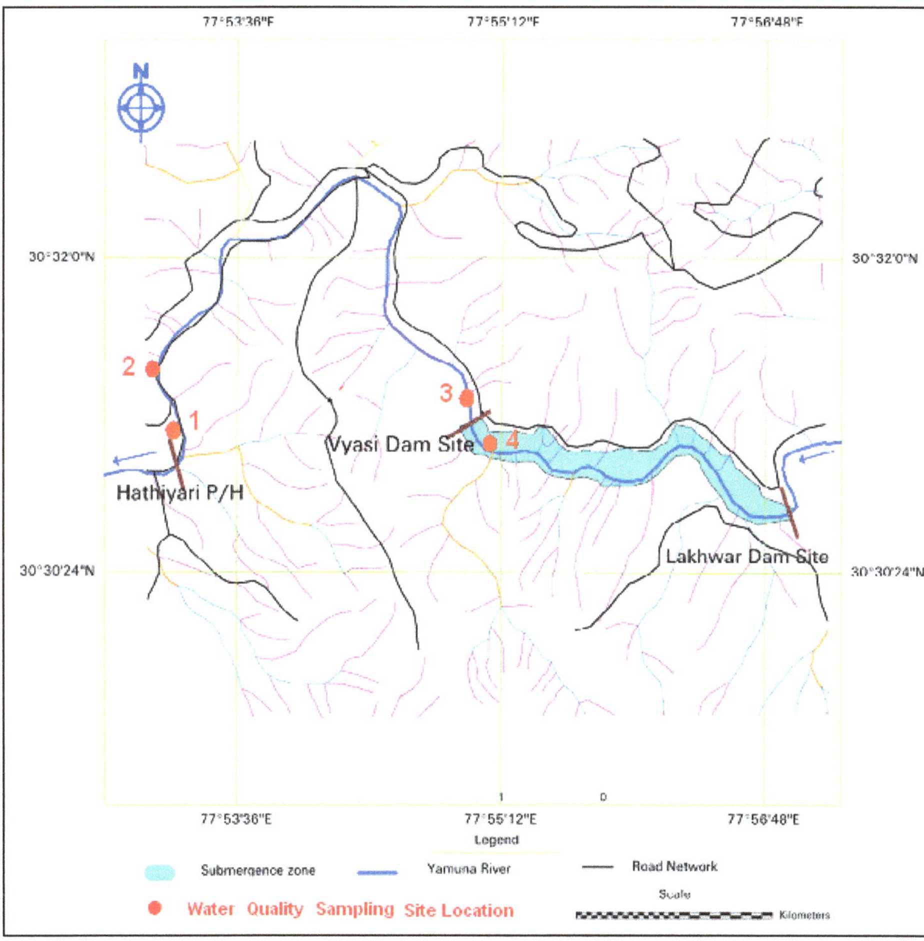

Figure 8.7: Drainage network map in the influence zone of Vyasi HEP and location map of water quality sampling sites.

Water Quality Sampling

With the help of digitized drainage network layer, the water sampling sites were marked considering the confluence of other small/big stream with the Yamuna within influence zone (Figure 8.7).

Roads and Settlements in the Project Area

The road and settlement location within the project area were digitized to create vector of roads network and point information on settlements location (Figure 8.8). This information was further overlaid on different other raster/vector layers and composed maps were used in baseline data collection and preparation of few management plans also. During field surveys for biodiversity assessment the road network laid on forest cover and land-use raster layers was used for spotting the targeted quadrate locations (as presented in Figure 8.6). Similarly, such other overlay maps were used in socio-economic and ethnographic surveys.

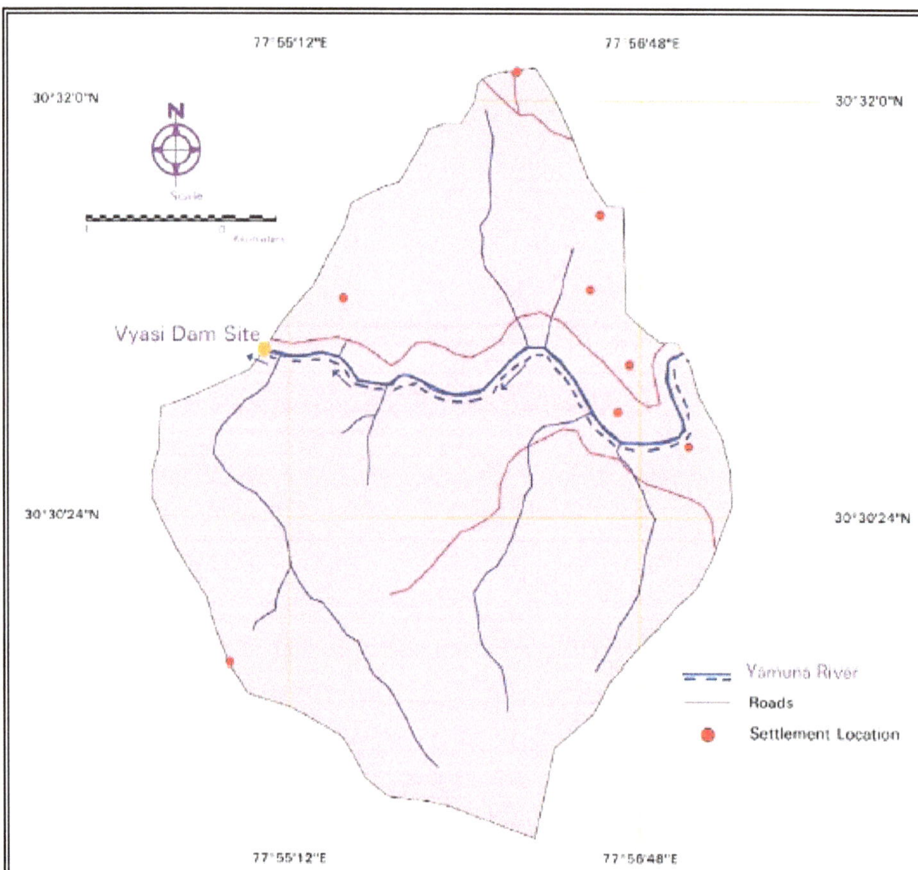

Figure 8.8: Roads and settlement map in the free draining catchment of Vyasi HEP.

RS-GIS in EMP Formulation

Catchment Area Treatment (CAT) Plan

The CAT envisages ensuring that the transport of sediment generated due to project induced as well as natural and human activities will be minimized up to an extent possible such that the life of reservoir in able to achieve the parameters of reservoir siltation rate considered in design. RS-GIS can serve many interests in catchment area treatment mainly for planning logistics such as identification of vulnerable areas and prioritization of watersheds. During last few decades RS-GIS approaches have gained importance in soil loss estimation, generating overlays and making site-specific decisions, etc. (Ahmet, *et al.*, 2007; Wu and Wang, 2007; Sekhar, *et al.*, 2002; Maitra, 2001; Hazarika, M.K. *et al.*, 1999; Garg and Seth, 2007).

RS data is being extensively used in watershed studies for deriving the parameters of the Sediment Yield Index (SYI) model to provide quantitative silt load estimates in watersheds (Kokh-Shesthra, 2002; AISLUS, 1991). Due to rapid development of hydropower projects in the Himalayan river valleys, there is an earnest need of effective CAT plans for the longevity of the reservoir of the project but as well as for the conservation of Himalayan natural resources and agriculture. This task can be made easy with support of advanced techniques like RS-GIS. Determination of erosion intensity unit is primarily based upon the integrated information on soil characteristics, physiography, slope and land-use/land-cover. The present study

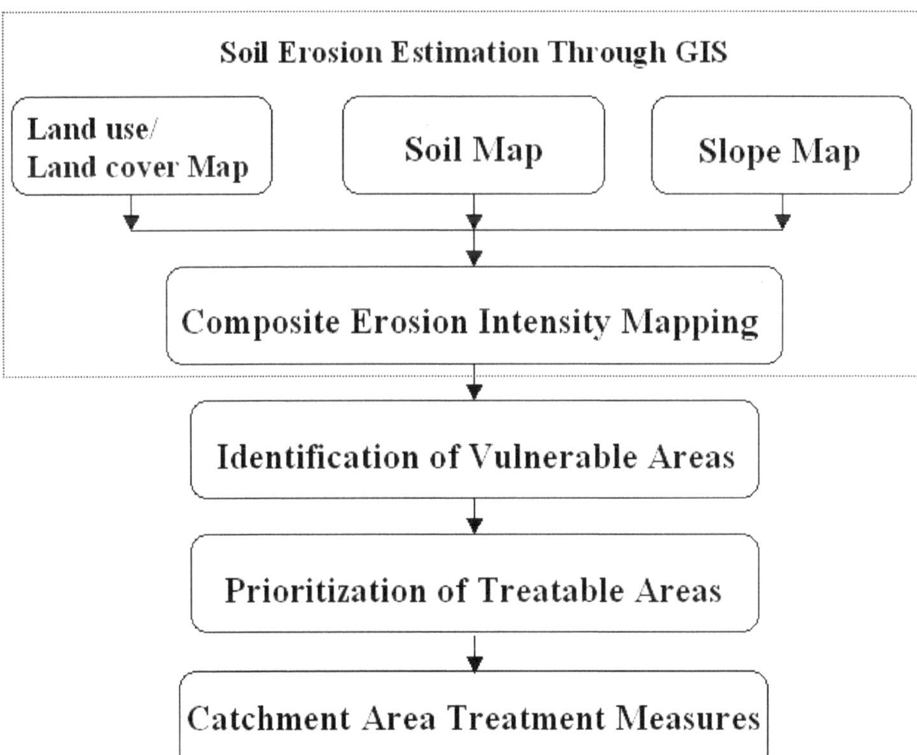

demonstrates the use of RS-GIS approaches for better CAT Planning. GIS tool, ERDAS IMAGINE was used for carrying out the entire exercise. The available data in the form of different thematic layers *viz.*, land-use/land-cover, slope and soil layers was classified using 'Knowledge Classifier' and a composite erosion intensity map was generated to work out the priority areas in different land-use classes. Soil and water conservation engineering knowledge base and logics were used to select the treatments in the vulnerable areas within the catchment. The flowchart below presents the outline of general procedure adopted for implementation of the approach.

Adopting the methodology outlined above in conjunction with the RS-GIS base layers, area falling under different erosion classes was as below:

Erosion Class		*Area (ha)*
Very Severe (E1)	:	2.46
Severe (E2)	:	184.46
Medium (E3)	:	1046.23
Low E4	:	84.00
Negligible (E5)	:	3.26
Water	:	40.96

To understand the land-use of areas vulnerable to soil erosion; results of composite erosion intensity map were superimposed on land-use classification (Figure 8.9). The free draining catchment was therefore evaluated for knowing the intensity of erosion in five land-use classes and accordingly the vulnerable areas belonging to very severe 'E1' and Severe 'E2' classes, were identified to be treated through various engineering and biological measures (Table 8.4).

Table 8.4: Vulnerable area requiring interventions based on erosion intensity classification in the free draining catchment of Vyasi HEP.

Sl.No.	*Land-Use Type*	*Very Severe (E1), ha*	*Severe (E2), ha*	*Total Treatable Area, ha*
1.	Agriculture	1.89	92.97	94.86
2.	Barren	0.17	41.66	41.83
3.	Scrub	0.40	28.91	29.31
4.	Open forest	0.00	20.91	20.91
5.	Dense forest	0.00	0.00	0.00
	Total	**2.46**	**184.46**	**186.91**

As per the need of treatment required under different land-use classes, suitable treatment models were proposed. Further, an additional exercise was done to distribute the suggested treatment measures for agricultural land at village scale. The overlays of village boundaries on erosion intensity map were made and village wise treatable area and subsequently the number of treatment measure were calculated.

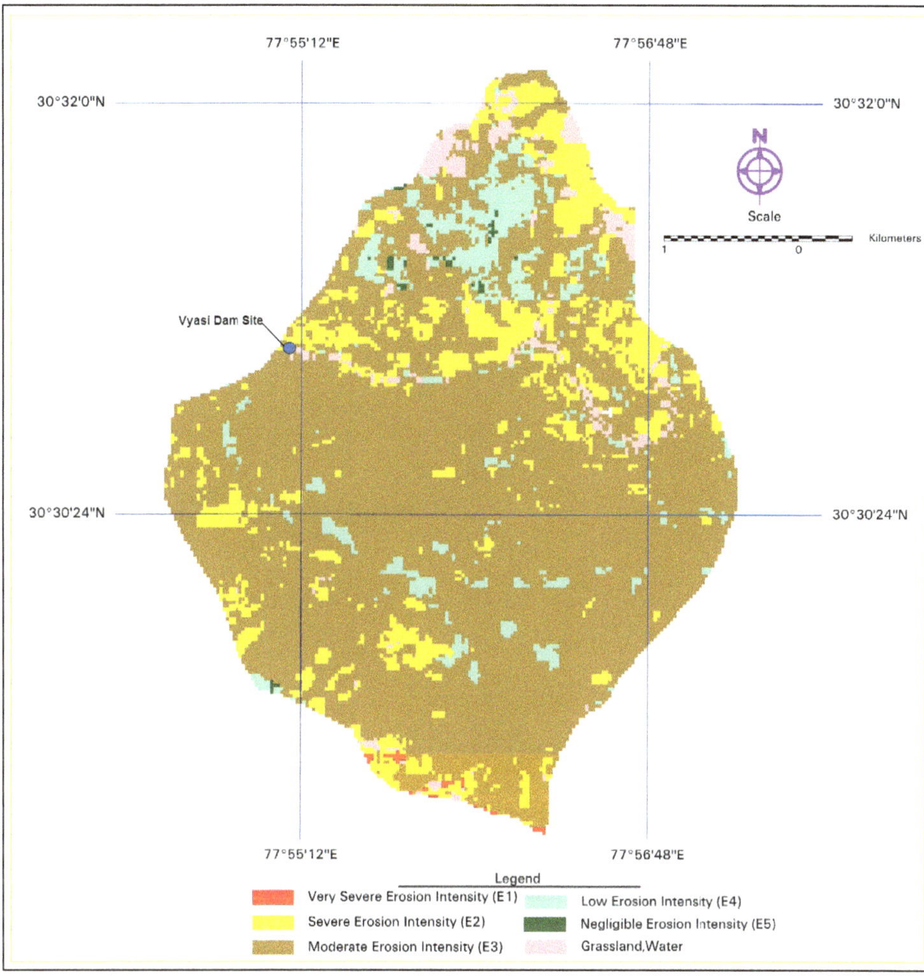

Figure 8.9: Composite erosion intensity map of the free draining catchment of Vyasi HEP.

Reservoir Rim Treatment Plan

As discussed earlier, construction of Vyasi dam will create a reservoir; and a separate management plan was suggested for the treatment of surrounding lip of the reservoir. The aim of the reservoir rim treatment is to check direct sedimentation in the reservoir, and to eradicate the probability of sliding of any material on critical slopes, landslides; stabilization of projects structures situated along the Reservoir Rim etc. After an extensive survey surrounding the reservoir rim, the zone of weaknesses (telus cones, stream network draining to the reservoir and the existing landslides) was delineated. With the help of digitized drainage map, it was seen that 22 streams would be directly draining into the proposed reservoir. For the treatment of these streams, it was proposed that just above the FRL level, a series of gabion check dams of sufficient dimensions should be constructed. Based on the slope map,

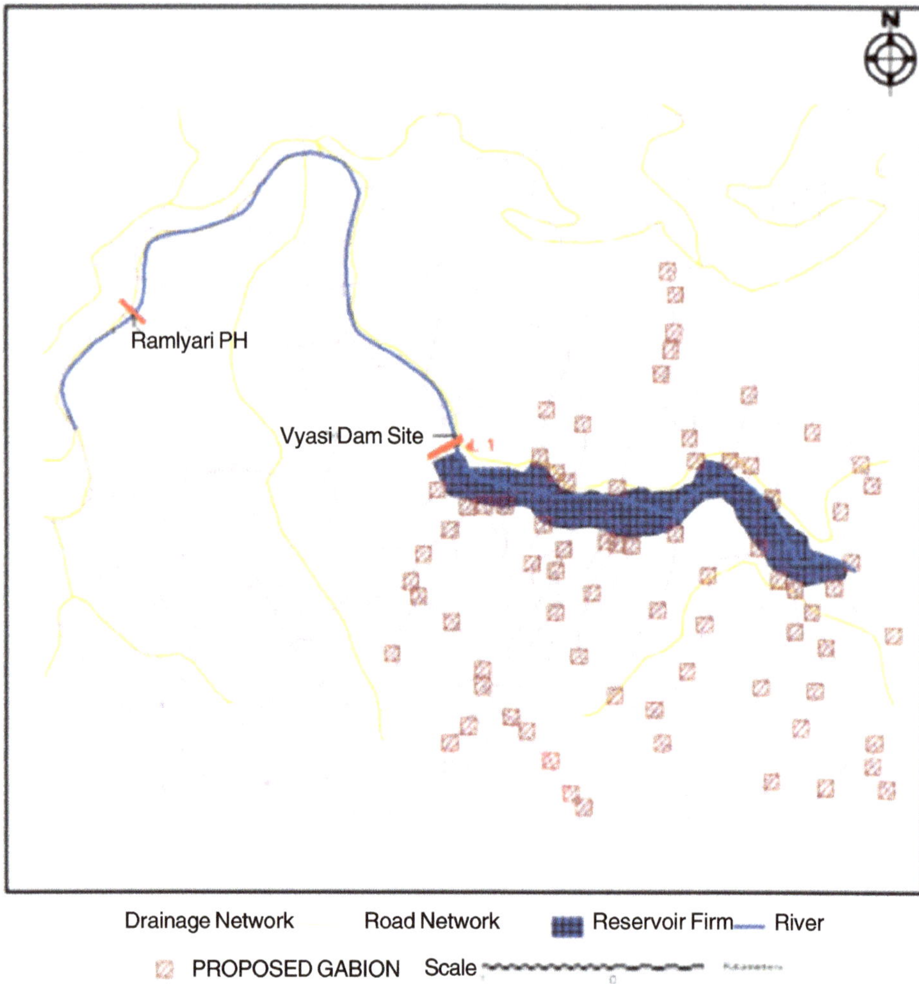

Figure 8.10: Location of proposed treatments for reservoir rim treatment in Vyasi HEP.

soil map and field survey of the area, appropriate locations for these structures were suggested. A map for the same is presented as Figure 8.10.

Disaster Management Plan: Flood Inundation Mapping

To assess the adverse effects of a dam failure, it is imperative to model a hypothetical scenario using available dam break modeling tools. A precision modeling of dam break floods can lead to formulation of proper emergency action plan to minimize flood impacts within the economic lifetime of the assets. Application of GIS techniques in integration of the results of hydrological modeling with mapping of the flood inundated areas can be a great asset in further minimizing the risk and likely damages. In the present study, dam break flow depths generated by the DAMBRK model under

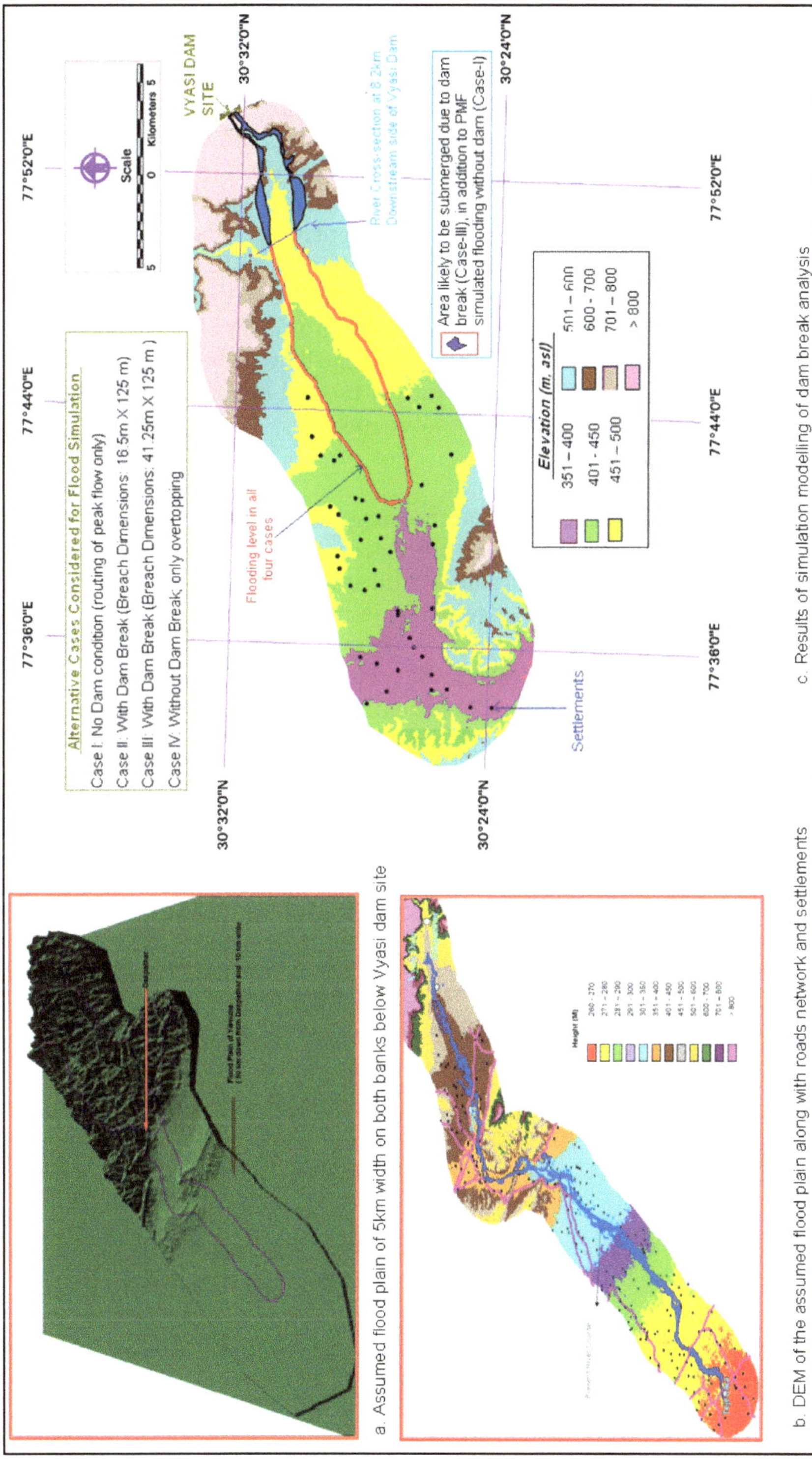

Figure 8.11: Flood inundation map in assumed floodplain of 10km width in downstream of Vyasi HEP.

various combinations of structural failure were subsequently plotted on Digital Elevation Model (DEM) of the downstream of project to map the likely affected area.

For defining the likely impact zone 5 km lateral stretch on both the banks of river Yamuna in the downstream of the proposed dam site was considered. On the base layer, human settlements, roads network, etc. were superimposed. In order to develop the DEM of the entire stretch, the contour map of the area was used and than in GIS platform TIN was prepared. Using these layers in conjunction with the results of dam break analysis, the flooding elevations of different reaches were superimposed on the DEM and inundation map is finally prepared (Figure 8.11). For the purpose of this analysis, additional area that is likely to be inundated in case of worst case of dam failure over and above the case of inundation as a result of flood occurrence without dam has been considered. The results indicate that, no human settlement is likely to be affected as a result of dam failure in this 8.2 km stretch; although an additional area of 4.69 sq.km. is likely to be submerged.

Conclusion

The present paper demonstrates that the integration of RS-GIS approaches in EIA studies can not only result in better and efficient baseline data generation, creation and compilation but it can also contribute significantly in formulating appropriate EMP. In the present era of raising sensitive environmental concerns, rapid advancements in RS-GIS data/techniques, can lead in better decision making. The researchers working on environmental related subject matters must proceed to take full advantage of RS-GIS techniques in enhancing capabilities, efficiency and creditability of EIA studies. Espousing these approaches is indeed beneficial for both, environmental as well RS-GIS sectors. The demonstration of such practical application will not only motivate environmentalists to follow these approaches but also gives clues to RS-GIS professionals to think for customized developments for environmental planning and decision making.

References

Ahmet I., F. T. lu, V. Uygur (2007). "Estimating spatial distribution of soil loss over Seyhan River Basin in Turkey", *Journal of Hydrology*, Vol. 336, pp. 30– 37.

AISLUS - All India Soil and Land-use Planning (1991). "Methodology for priority delineation survey", *Ministry of Agriculture, Government of India*, New Delhi, India.

Antunes, P., R. Santos, L. Joao, P. Goncalves, and N. Videira (1996). "A GIS Based Decision Making System for Environmental Impact Assessment", in: Proceedings of the International IAIA Conference, 17-23 June 1996, Estoril, Portugal, pp. 451-456.

Berry, J. K. (1996). "Cartographic Modeling: The Analytical Capabilities of GIS", in: M. F. Goodchild, and B. O. Parks (Eds), *Environmental Modeling with GIS*, Oxford University Press, pp 59-74.

DPR (2007). "Detailed Project Report for Lakhwar-Vyasi Hydroelectric project", National Hydroelectric Power Corporation (NHPC), India.

Eedy, W. (1995). "The Use of GIS in Environmental Assessment", *Impact Assessment (IAPA)*, Vol. 13, No. 4, pp. 199-206.

Garg V. and R. Seth, (2007), "Remote Sensing and GIS in identifying areas vulnerable to soil erosion", available at: http: //www.gisdevelopment.net/application/natural_hazards/landslides/index.html (accessed 15 September 2008).

GBPIHED (2007). "Comprehensive Environmental Impact Assessment for Lakhwar hydroelectric project", prepared for National Hydroelectric Power Corporation, India.

Goodchild, M.F. (1993). "A Perspective on the State of Environmental Simulation Modeling", in: M. F. Goodchild and B. O. Parks (Eds), *Environmental Modeling with GIS*, Oxford University Press, pp 16-30.

Haklay, M., E. Feitelson, and Y. Doytsher (1998). "The Potential of a GIS-Based Scoping Sytem: An Israeli Proposal and Case Study", *Environmental Impact Assessment Review*, Vol. 18, Issue 5, 1998, Pages 439-459.

Hazarika, M.K. and K. Honda (1999). "Estimation of soil erosion using remote sensing and GIS data, its valuation and economic implication on agricultural production", International Soil Conservation Organization Meeting, 24-29 a Purdue and the USDA-ARS National Soil Erosion Research Laboratory.

Joao, E. and A. Fonseca (1996). "The Role of GIS in improving Environmental assessment Effectiveness: Theory versus Practice", *Impact Assessment, International Association for Impact Assessment*, Vol. 14, December, pp. 371-385.

Kokh-shesthra M. (2002). "Soil erosion modeling using Remote Sensing and GIS: a case study of Jhikhu Khola watershed, Nepal", in: *23rd Asian Conference on Remote Sensing*, Available at: www.gisdevelopment.net/aars/acrs/2002/sol/244.pdf (accessed 2 August 2007).

Maitra Shibani (2001). "Environmental Impact Assessment for dam construction using GIS/Remote Sensing", in: ESRI Proceedings, International User Conference, available at: http: //gis.esri.com/library/userconf/proc01/professional/papers/pap1043.htm (accessed 13 December 2008).

NBSS and LUP (2002). "National Bureau of Soil Survey and Land-use Planning Soils of India", Indian Council of Agricultural Research, New Delhi, India.

Nyerges, L. T. (1993). "Understanding the Scope of GIS: Its Relationship to Environmental Modeling", in M. F. Goodchild and B. O. Parks (Eds), *Environmental Modeling with GIS*, Oxford University Press, 75-91.

Sekhar, K. R. and B. V. Rao (2002). "Evaluation of sediment yield by using remote sensing and GIS: a case study from the Phulang Vagu watershed, Nizamabad District (AP), India", *International Journal of Remote Sensing*, Volume http: //www.informaworld.com/smpp/title~content=t713722504~db=all~tab=issueslist~branches=23 - v2323, Issue 20, pp 4499 – 4509.

Treweek, J. and N. Veitch (1996). "The potential application of GIS and remotely sensed data to the ecological assessment of proposed new road schemes", *Global Ecology and Biogeography Letters*, 5 (4/5), 249-257.

Wu Q. and M. Wang (2007). "A framework for risk assessment on soil erosion by water using an integrated and systematic approach", *Journal of Hydrology*, Vol. 337, pp. 11– 21.

9

Groundwater Development Using Remote Sensing and GIS

Y. Rama Mohan, Avinash Kumar & Anil K. Gupta

INTRODUCTION

The groundwater levels are decreasing over the years due to decline in annual rainfall, climatic changes, rapid urbanization, industrialisation, increasing runoff, agricultural activities and deforestation. Hence, it is necessary to increase groundwater levels in the perspective of groundwater use for future demands. Satellite images are increasingly used in groundwater exploration to identify the various ground features, which may serve as either direct or indirect indicators of groundwater potential (Bahuguna *et al.*, 2003; Das *et al.*, 1997) as well as identifying landform features, drainage pattern and geomorphic indicators for location of recharge and discharge areas (Sharma and Jugran, 1992; Chatterjee and Bhattacharya, 1995; Tiwari and Rai, 1996; Ravindran, 1997; Srivastava and Bhattacharya, 2000; Sarkar *et al.*, 2001; Khan and Moharana, 2002; Sankar, 2002). Hydrogeomorphological mapping using remote sensing data has been used conventionally for delineating groundwater prospect zones in many regions. In the complex terrains comprising fluvial, denudational and structural geomorphic units, the intricate relationship among the various terrain parameters, it is difficult to evaluate the groundwater potential. The GIS has emerged as a powerful tool in analyzing and quantifying such multivariate aspects of groundwater occurrence. It is very helpful in delineation of groundwater prospects and deficit zones (Carver, 1991; Goyal *et al.*, 1999). Keeping this in view, the present study aims at utilizing the advanced techniques of remote sensing and GIS in groundwater potential mapping and to suggest remedial measures to improve groundwater potential in the arid and semi-arid regions.

Study Area

The study area comprises part of Hyderabad and Mahabubnagar districts of Andhra Pradesh bounded by 17°0' to 17°45'N latitude and 78°30' to 78°45'E longitude (Figure 9.1). It is located in the Survey of India (SOI) topographical map 56K/12 and occupies an area of 217 km^2. The average annual rainfall in the study area is 890 mm. The mean daily temperature varies from 30°C to 36°C from April to June and from 20°C to 24°C in the months of December and January. The mean maximum temperature ranges between 40°C and 47°C in May.

Methodology

In the present study, evaluation of groundwater potential has been attempted by preparing thematic maps namely, hydrogeomorphology, geology, drainage, lineament, slope, and relief by integrating them in a GIS domain. Thematic maps were prepared by using digitally geocoaded IRS-1D, LISS-III and PAN images (December 2004), SOI topographic map (56K/12, on 1:50,000 scale) and geological map compiled by the Geological Survey of India. The geo-referenced LISS and PAN images were further merged, using ERDAS v.9 software, to obtain a fused high resolution (5.6 m) false colour composite. Various digital image processing techniques, *viz.*, principal component analyses, band ratio, band multiplication, enhancements, filtering and supervised classification (Jensen, 1986; Mather, 1987) have been used to process and analyse the digital data.

Soil map of the study area was prepared using the satellite image, National Bureau of Soil Survey map and Land-use Planning (ICAR) map. Geomorphological landforms were studied and classified using remote sensing data followed by ground-truth observations. The groundwater potential map was prepared based on the analysis of various themes such as geomorphology, land-use/cover (LU/LC), lineament, intersection points, drainage pattern and lithological evidences, besides the collateral data obtained from State Groundwater Board, and field surveys by using converging evidence concept. The groundwater prospects map was generated by integrating the geomorphology, structure, lithology and drainage (hydrological) maps. A probability-weighted approach has been applied during overlay analysis in ArcGIS environment. The overlay analysis allows a linear combination of weights of each thematic map with the individual capability value with respect to groundwater potential (NRSA, 2000).

Creation of Thematic Maps and Spatial Database

Various thematic maps needed for evaluating groundwater potential were prepared using topomap, remote sensing and field survey data which are automated by digitization process using AutoCAD software. These maps were prepared on 1:50,000 scale and the attributes of entities are shown by different symbols/colours. These digitized data were exported into ArcInfo and ArcView file formats (*i.e.*, coverage and shapefile, respectively), and digital database were created for analyses (Mark, 1994). The system pioneered by United States Geological Survey (USGS) modified by

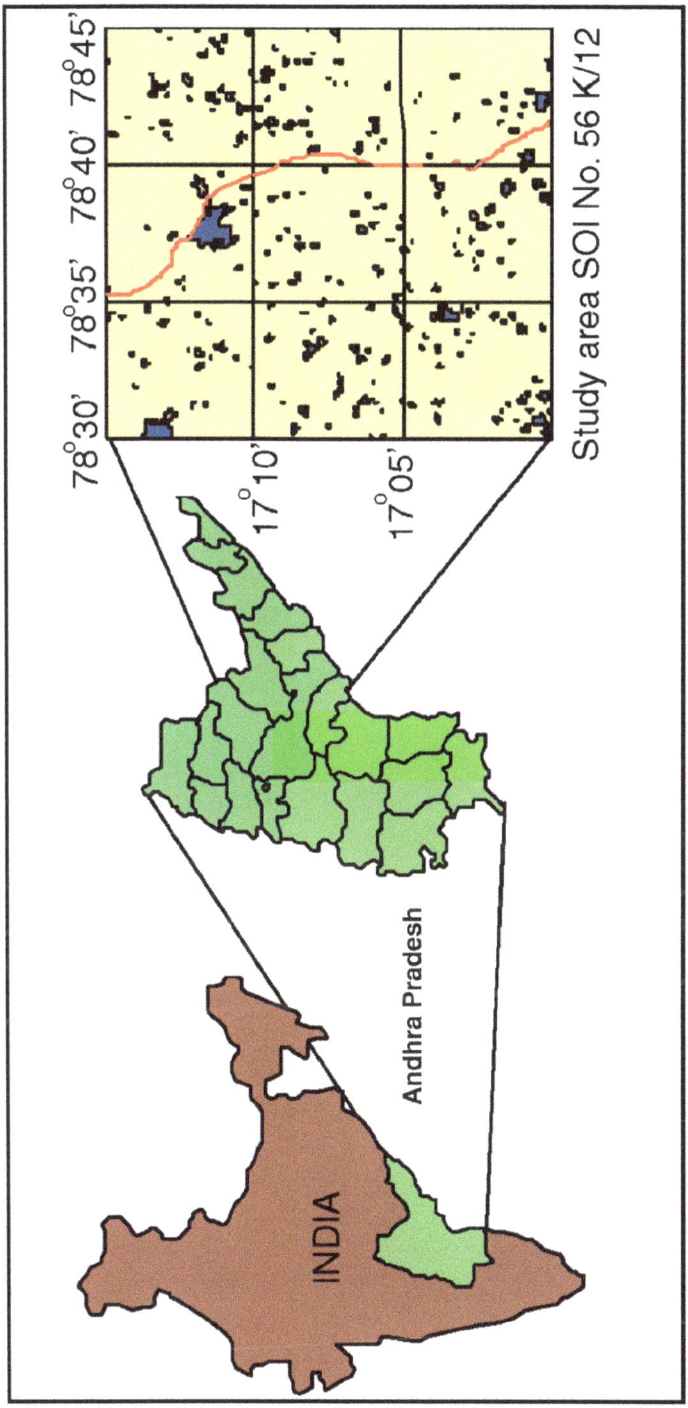

Figure 9.1: Location map showing the study area.

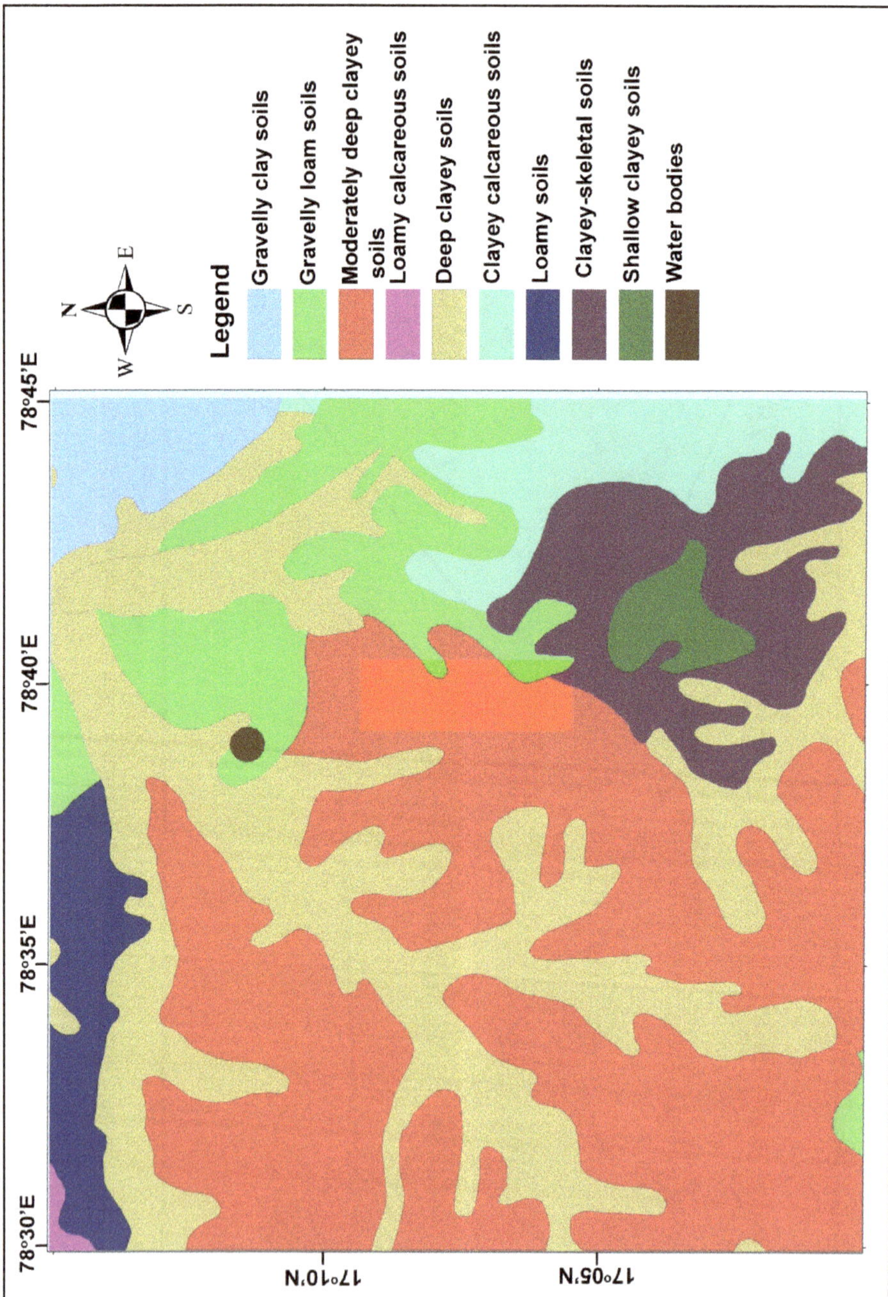

Figure 9.2: Soil map of the study area showing the distribution of various types.

National Remote Sensing Agency (NRSA) according to Indian conditions is used in this study for generation of thematic maps which include LU/LC, geomorphology, geology, soil, infiltration rate, structures, lithology, landforms, slope, lineament density, drainage density and drainage frequency and water table fluctuation. The attribute database includes the collateral (groundwater level, demography etc.) data collected during field survey and from government organizations. These attribute data are further integrated with the spatial data in a GIS domain.

Geology and Structure

Geological mapping is a two dimensional representation of patterns of lithological boundaries, in relation to topography and structure (Bhattacharya, 1973). The area constitutes mainly a granitic terrain (Pink-grey) exposing a variety of Archaean granitites of peninsular gneissic complex (PGC) and schistose (older metamorphic) rocks undulating with scattered denudation hills and ridges. They are intruded by basic dykes (Proterozoic) and covered locally by the Deccan traps (Upper Cretaceous to lower Eocene). The PGC of the area are classified into three main suites based on the petro-chemical characteristics and field relation. They are: i) Tonalite-trondhjemite-quartz chlorite (TT), ii) Adanellite-granodiorite-quartzdiorite (AG), and iii) Granite alkali feldspar granite.

Lineaments are commonly associated with dislocation and deformations, which are categorised into inferred (areas with high drainage density) and conformed (areas with low drainage density) lineaments. The study area is dominant by conformed lineaments, whereas in the regions where inferred lineaments are found can be predicted as good groundwater potentials zones. Lineaments, joints and faults zones are hydro-geologically very important to provide the pathways for groundwater movement (Sankar *et al.*, 1996). The identified lineaments are trending in north-south, northeast-southwest, northwest-southeast and east-west and few of them intersect each other. The majority of the lineaments correspond to either dyke ridges or stream channels, which are important for occurrence and movement of groundwater.

Soil

Based on the particle size and water holding capacity, nine types of soil were identified in the study area (Figure 9.2). The identified soil types have been broadly grouped into gravelly clayey soil, gravelly loamy soil, moderately deep clayey soil, loamy calcareous soil, deep clayey soil, clayey calcareous soil, loamy soil, clayey skeletal soil, and shallow clayey soil. Each one of these soils is further subdivided into soils with very low water capacity, low water capacity, medium water capacity, high water capacity and very high water capacity.

Slope

The altitude varies from 500 m in the west to 400 m in the east with a maximum elevation of 697 m in the SE region and upto 8 different slope classes are observed in the study area (Figure 9.3). Nearly flat level slope class covers 88 per cent of the total study area, whereas the distance between the successive contours is more. About 10 per cent of the area falls within gently sloping, 2 per cent moderate sloping, 0.35 per cent moderately steep sloping, 0.06 per cent steep sloping, and 0.08 per cent very steep sloping classes (Figure 9.3).

Groundwater Infiltration

The infiltration rate plays an important role in determining the level of groundwater contamination. Infiltration rate is determined on the basis of soil types, their thickness and weathering status of rocks. Based on the infiltration rate, the study area has been divided into high, medium, and low infiltration zones. Low infiltration rate has been observed in impermeable, compact, sheet rock and urban areas. Whereas, high infiltration rate has been observed wherever, thickness of topsoil is moderate and the slope levels are conducive for vertical percolation as well as areas along river and surface water bodies.

Geomorphology

Geomorphologic landforms are divided into i) pediplain with moderate weathering (PPM) (37.9 per cent), ii) pediplain with shallow weathering (PPS) (8.6 per cent), iii) pediment (PD) (20.9 per cent), iv) pediment inselberg complex (PIC) (27.1 per cent), v) dyke (2.2 per cent), and vi) valley fill (0.8 per cent) (Figure 9.4). Due to

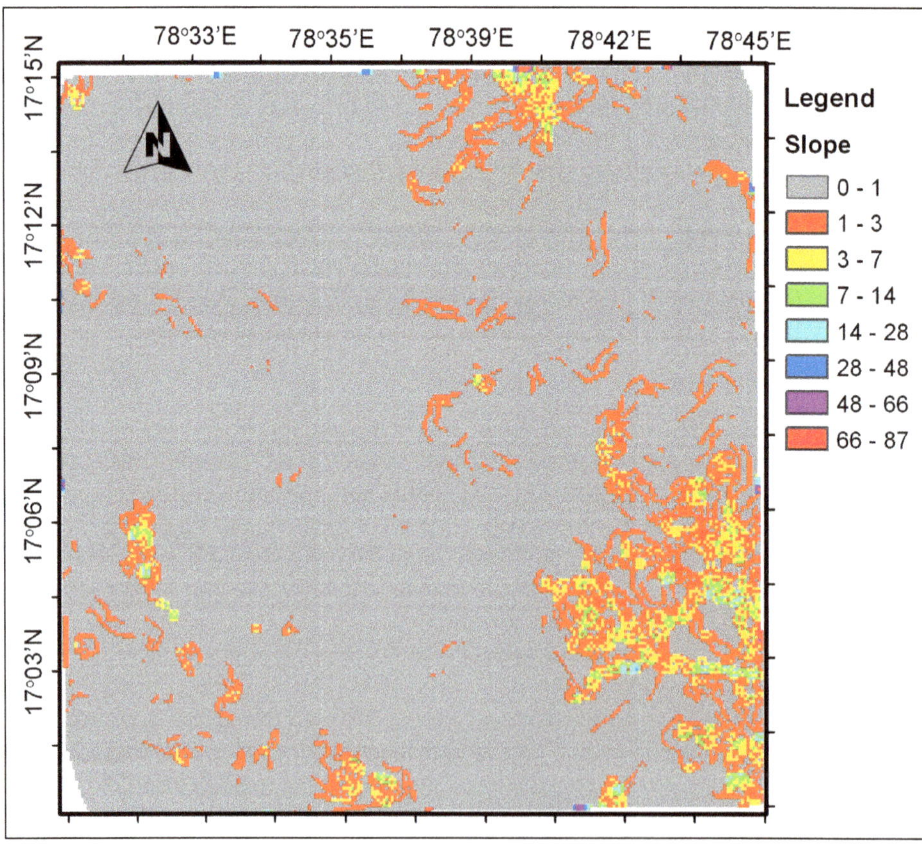

Figure 9.3: Slope map of the study area.

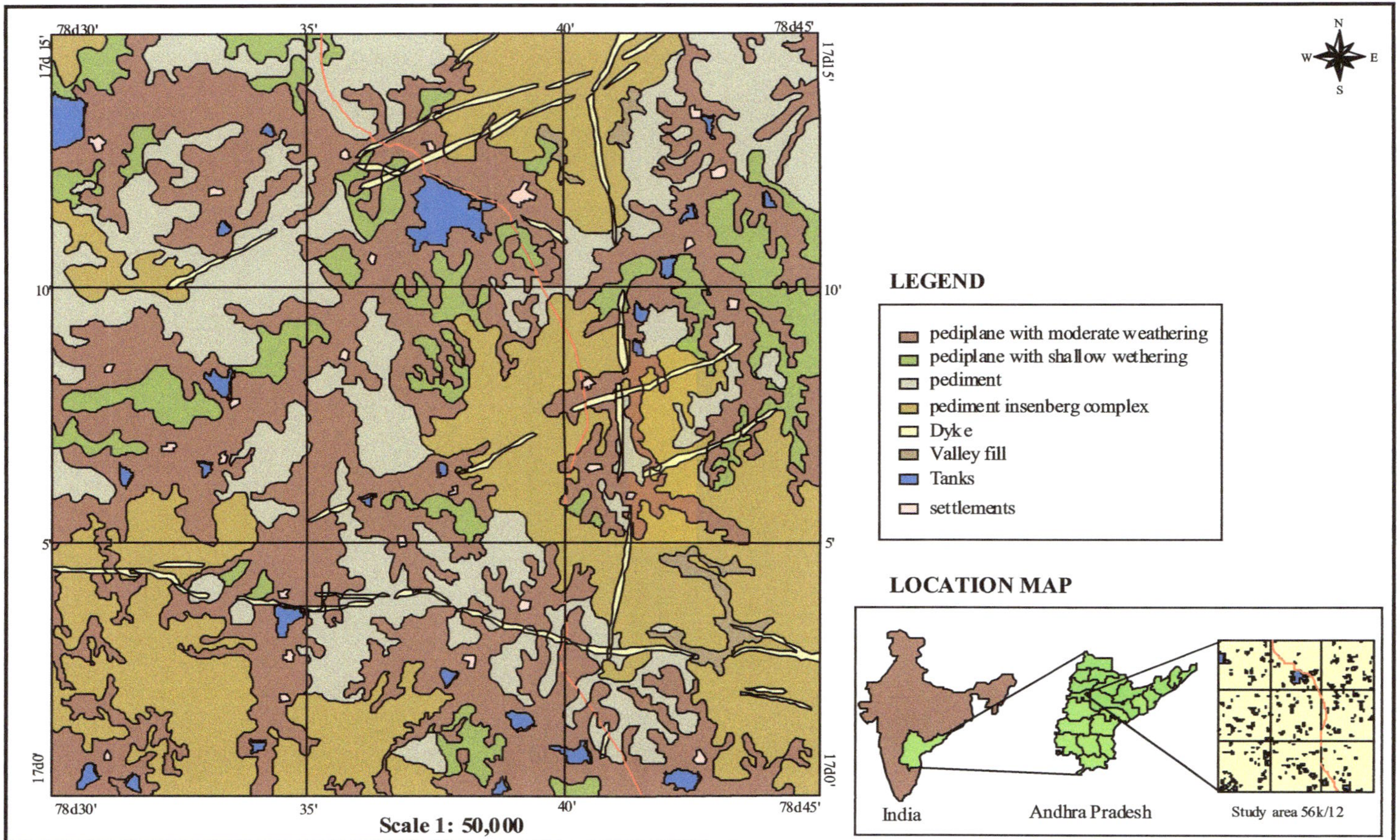

Figure 9.4: The hydrogeomorphic map showing various geomorphologic landforms.

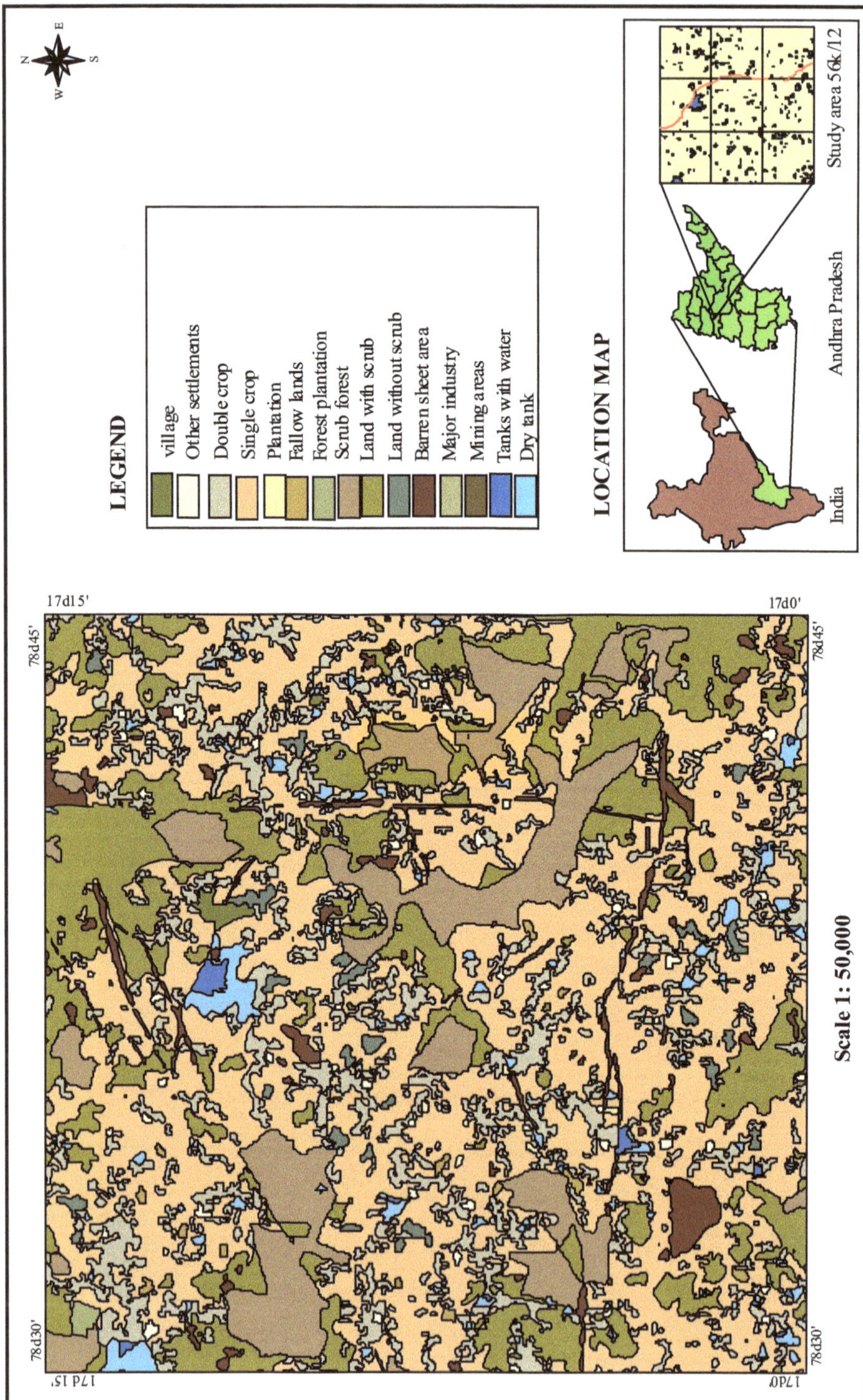

Figure 9.5: The map showing the land use/land cover distribution of the study area.

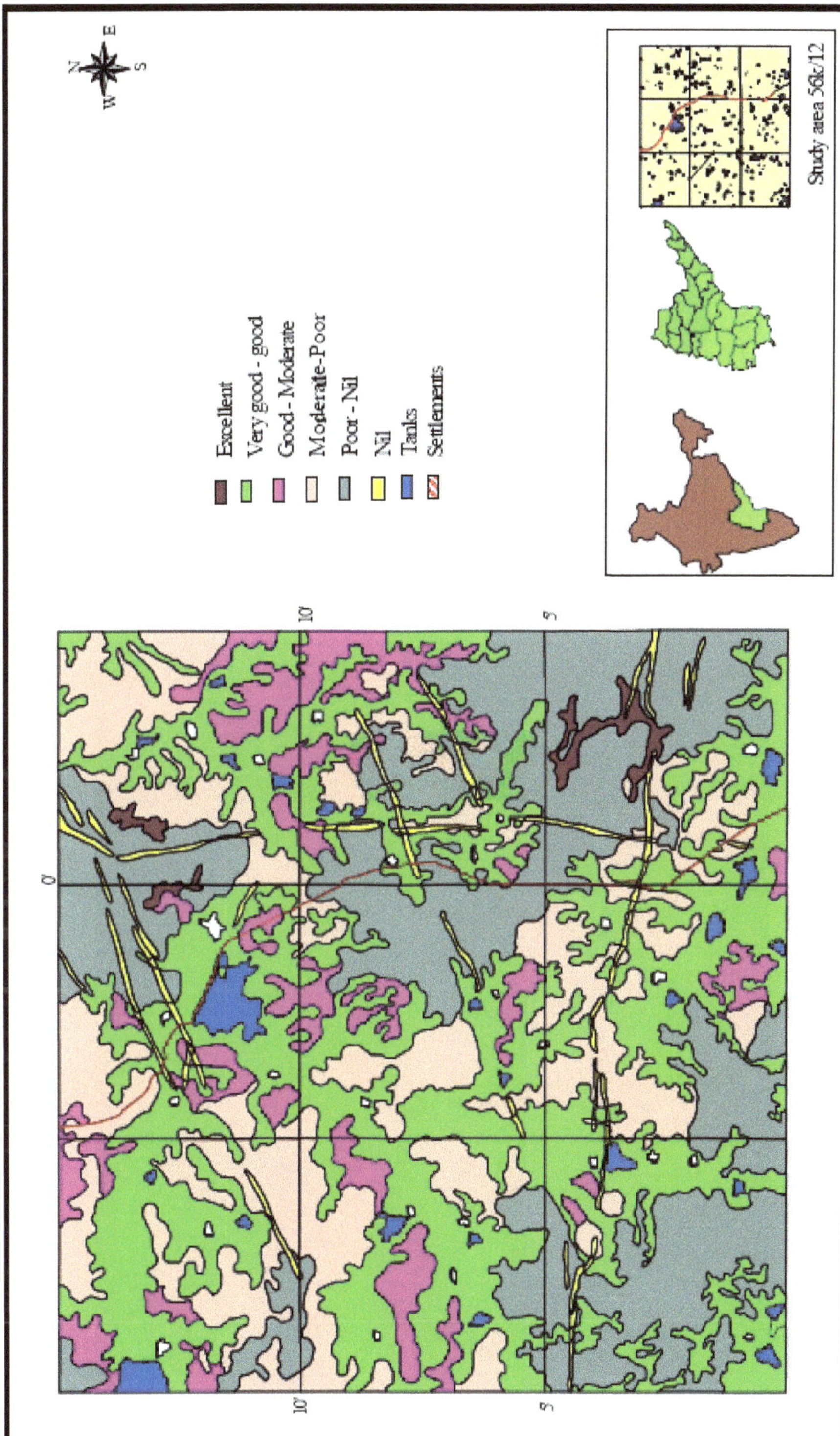

Figure 9.6: Groundwater potential map of the study area.

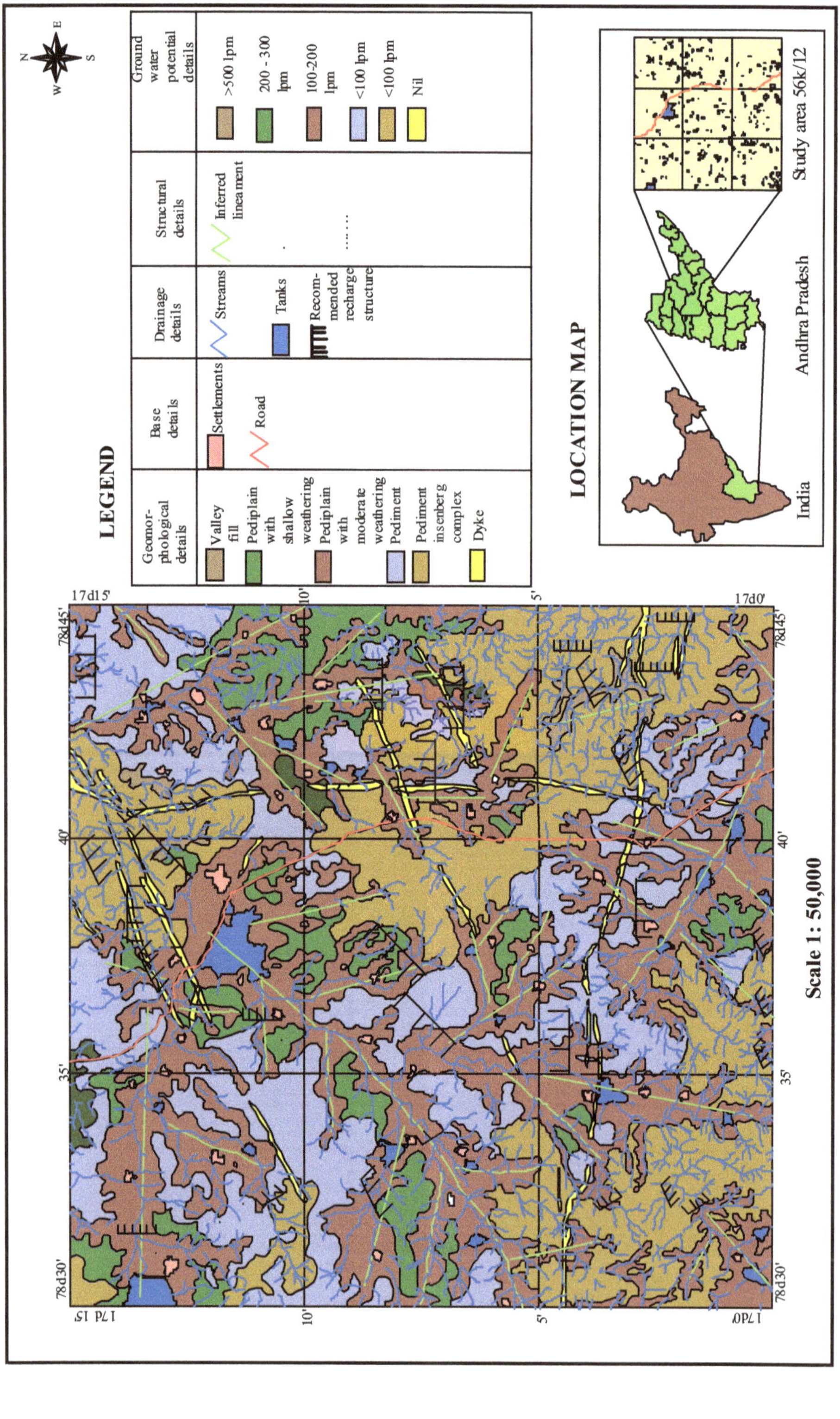

Figure 9.7: Groundwater prospects map of the study area.

variation in thickness of weathered materials, infiltration rate of PPM is high (10-20 m) compare to PPS (0-10 m). Hence, high groundwater potentials can be predicted at PPM than PPS (NRSA, 2000). Moderate to poor yields of groundwater can be predicted at pediments based on the rock formation and structures. In pediment inselberg complex, inselbergs form run-off zones and pediments contribute for limited to moderate recharge. Whereas, dyke acts as a barrier for streams, hence moderate groundwater potential can be obtained on one side of the dyke but the other side may be dry.

Land-Use/Land-Cover

The different LU/LC categories such as, agriculture, water bodies, forest, wastelands, built-up land and others have been identified and mapped. From the satellite data, agriculture area (62 per cent) could be clearly delineated into three categories namely, single crop, double crops, and plantation crops. Though single crop has been observed at various parts of the study area, double crops have been observed on banks of the Musi River and its major streams. Plantation crops are also observed at some places. About 3 per cent of the study area is occupied by water bodies - tanks with water and without water. Forest - including scrub forest and forest plantation - covers ~10 per cent of the study area. About 23 per cent of study area is under wastelands, which is further classified into land with scrub, land without scrub, fallow land and barren sheet area. Built-up lands like villages and towns comprise about 1.5 per cent, whereas, only 0.8 per cent is occupied by mining and major industries (Figure 9.5). Within the LU/LC categories, it has been noticed that groundwater potentials are low at built-up land and barren sheet rock areas due to less infiltration rates, whereas, the areas near to water bodies have high and moderate groundwater potentials.

Groundwater Potential

Groundwater potential map prepared using various thematic maps by converging evidence concept, collateral data and evidences collected during field surveys. Based on the availability and quantum of groundwater, the whole study area is divided into five zones *viz.*, excellent, very good to good, good to moderate, poor-nil and nil (Figure 9.6). The excellent and very good potential areas represent adequate groundwater resources, the moderate potential zones represent areas depleted due to excess withdrawals of groundwater, and poor potential zones represent very scares groundwater sources. The excellent groundwater potential zones are observed in valley fill regions of the study area. The geomorphic unit, pediplain with moderate weathering (PPM) surrounding major tanks indicates very good groundwater potential, whereas pediplain with shallow weathering (PPS) indicates good to moderate groundwater potential. The areas cover under other geomorphic units like pediments, inselbergs and dykes are found to be poor to nil in groundwater potential.

Groundwater Prospects

Groundwater prospects map of the area is prepared by following the three steps: i) lithological-geomorphic units were prepared by superimposition of the geomorphological and lithological maps, ii) superimposing the geological structural map on the integrated lithological-geomorphic units, and iii) the hydrological information is transferred on to the integrated lithological-geomorphic-structure map. In addition to the above, metal roads and settlements are also superimposed on the integrated map and a final groundwater prospects map is obtained (Figure 9.7).

The groundwater potential ranges from excellent (>500 lpm), very good to good (200-300 lpm), good to moderate (100-200 lpm), moderate to poor (<100 lpm), poor to nil (<50 lpm) and nil (in the rocky out crops). Based on the above integrated information, suitable locations are identified for construction of recharging structures like check dams and percolation tanks, in order to increase the groundwater potential.

Conclusion

The study and environs exhibit diverse hydrogeomorphological conditions where the groundwater regime is controlled by a variety of terrain parameters primarily the landforms, geology and slope. Groundwater prospects zonation based on geoinformatics approach clearly indicated that combination of alluvial plains, coarse drainage density and gentle slope areas are the favourable terrain conditions for having good groundwater potential. Structural and denudational hills, fine drainage density and high slope areas are have poor prospects. In the present study, the excellent (>500 lpm) groundwater potential zone is observed in valley fill areas; very good to good (200-300 lpm) at areas of pediplain with moderate weathering (PPM) surrounded by major tanks. Pediplain with shallow weathering (PPS) indicates good to moderate (100-200 lpm) potential zone. Whereas, groundwater potential is noticed to be poor to nil (<50 lpm) at those places where other geomorphic units like pediments, inselbergs and dykes are found. In order to increase the groundwater levels, suitable locations can be identified for construction of artificial recharging structures.

Recommendations

- ☆ To improve the groundwater potential and avoid its further depletions, several artificial recharge structures like roof top rainwater harvesting, pits and scavenger wells should be constructed in buildings.
- ☆ Regulation and control of groundwater development is required to be made operational for not only protecting the environment but also to ensure equity in sharing groundwater.
- ☆ Provision of drainage, controlled irrigation, off season tillage, use of acid forming fertilizers to affect toxicity due to high lime content.
- ☆ Systematic land development, choice of crops to suit root zone limitations, irrigation scheduling.
- ☆ Land shaping and land leveling, provision of effective internal drains.

Acknowledgements

The authors thank the National Bureau of Soil Survey, Nagpur and State Groundwater Board, Hyderabad for providing the data pertaining to soil and groundwater potential respectively. The authors are grateful to Dr. K.S. Jayappa,

Department of Marine Geology, Mangalore University for his valuable suggestions and discussion. Avinash Kumar thanks the Council for Scientific and Industrial Research (CSIR), India, for providing the Senior Research Fellowship.

References

Bahuguna, I.M., S. Nayak, V. Tamilarsan and J. Moses (2003). "Groundwater Prospective Zones in Basaltic Terrain using Remote Sensing", *Journal of the Indian Society of Remote Sensing*, Vol.31 No.2, pp.107–118.

Bhattacharya (1973). *Elements of Geological Map Reading and Interpretation*, Orient Longman Limited, Calcutta.

Carver, S. (1991). "Integrating Multi-Criteria Evaluation with Geographic Information Systems", *International Journal of Geographical Information Science*, Vol.5, pp.321–339.

Chatterjee, R.S. and Bhattacharya, A.K. (1995). "Delineation of the Drainage Pattern of a Coal Basin Related Inference using Satellite Remote Sensing Techniques", *Asia Pacific Remote Sensing Journal*, Vol.1, pp.107-114.

Das, S., S.C. Behra, A. Kar, P. Narendra and N.S. Guha (1997). "Hydrogeomorphological Mapping in Groundwater Exploration using Remotely Sensed Data – Case Study in Keonjhar District in Orissa", *Journal of the Indian Society of Remote Sensing*, Vol.25 No.4, pp.247–259.

Goyal, S., R.S. Bhardwaj and D.K. Jugran (1999). "Multicriteria Analysis using GIS for Groundwater Resources Evaluation in Rawasen and Pili Watershed, U.P.", *Proceedings Map India 99*, New Delhi, India.

Jensen, J.R. (1986). *Introductory Digital Image Processing*, Prentice Halls, Englewood Cliffs, NJ, P339.

Khan, M.A. and E.C. Moharana (2002). "Use of Remote Sensing and Geographical Information System in the Delineation and Characterization of Groundwater Prospect Zones", *Journal of the Indian Society of Remote Sensing*, Vol.31 No.3, pp.131-141.

Mark, D.M. (1994). "Determining Spatial Relations Between Lines and Regions in ARC/INFO using the 9-Intersection Model", *Proceedings 14th Annual ESRI User Conference*, pp.1207-1213.

Mather, P.M. (1987). *Computer Processing of Remotely Sensed Images: An Introduction*, John Wiley and Sons, NY, P352.

NRSA (2000). "Technical Guidelines for Preparation of Groundwater Prospects Map", Rajeev Gandhi National Drinking Water Mission, National Remote Sensing Agency, Hyderabad.

Ravindran, K.V. (1997). "Drainage Morphometry Analysis and its Correlation with Geology, Geomorphology and Groundwater Prospects in Zuvari basin, South Goa, using Remote Sensing and GIS", *Proceedings of National Symposim on Remote Sensing for Natural Resource with Special Emphasis on Water Management*, pp.270-296.

Sankar, K. (2002). "Evaluation of Groundwater Potential Zones using Remote Sensing Data in Upper Vaigai River Basin, Tamil Nadu, India", *Journal of the Indian Society of Remote Sensing*, Vol.30 No.3, pp.119-129.

Sankar, K., M.S. Jegatheesan and Balasubramaniam, A. (1996). "Geoelectrical Resistivity Studies in the Kanyakumari District, Tamil Nadu", *Journal of Applied Hydrology*, Vol.9 No. 1 and 2, pp.83-90.

Sarkar, B.C., B.S. Deota, P.L.N. Raju and D.K. Jugran (2001). "A Geographic Information System Approach to Evaluation of Groundwater Potentiality of Shamri Micro-Watershed in the Shimla Taluk, Himachal Pradesh", *Journal of the Indian Society of Remote Sensing*, Vol.29 No.3, pp.151-164.

Sharma, D. and D.K. Jugran (1992). "Hydromorphogeological Studies around Pinjaur-Kala Amb area, Ambala District (Haryana), and Sirmur district (Himanchal Pradesh)", *Journal of the Indian Society of Remote Sensing*, Vol.20 No.4, pp.281-286.

Srivastava, P.K. and A.K. Bhattacharya (2000). "Delineation of Groundwater Potential Zones in a Hard Rock Terrain of Bargarh District, Orissa using IRS Data", *Journal of the Indian Society of Remote Sensing*, Vol.28 Nos.2 and 3, pp.129-140.

Tiwari, A. and B. Rai (1996). "Hydrogeological Mapping for Groundwater Prospecting using Landsat-MSS Images - A Case Study of part of Dhanbad District, Bihar", *Journal of the Indian Society of Remote Sensing*, Vol.24 No.4, pp.281-286.

10

Hydraulic Simulation of Groundwater Flow

V. Ramani Bai, Sumiani Yusoff & S. Mohan

INTRODUCTION

The most formidable groundwater challenge is to attain the sustainable use and management of groundwater in vast and growing regions where the resource is under threat. The over-depletion of groundwater is becoming a major problem in the North China plains, in western and peninsular India and in urban areas across Asia. Over 25 per cent of India's harvest is in areas where groundwater is being seriously overexploited. Here the challenge is to sustainable development and management this potential, for the benefit of the poor people. The objective of this research work is to assess the groundwater potential in the Vaigai basin, to estimate the aquifer parameters and to compare the estimation of groundwater balance by different methods and to use most reliable result for the development of the model. These objectives will be helpful particularly in creating a detailed picture of groundwater situations and use patterns.

This also includes collection, synthesizing past and current scientific work to re-examine important technical and behavioral relationships related to groundwater. To formulate a realistic assessment of opportunities and threats. The Vaigai basin shown in Figure 10.1 (indicating Theissen polygons) in Tamil Nadu is chosen for this study. It has an area of 7000 sq.km where current (and projected) supplies of surface and groundwater are not deemed sufficient to meet current (and projected) needs. In 1985 a tunnel diverted waters from the Periyar river in Kerala under a

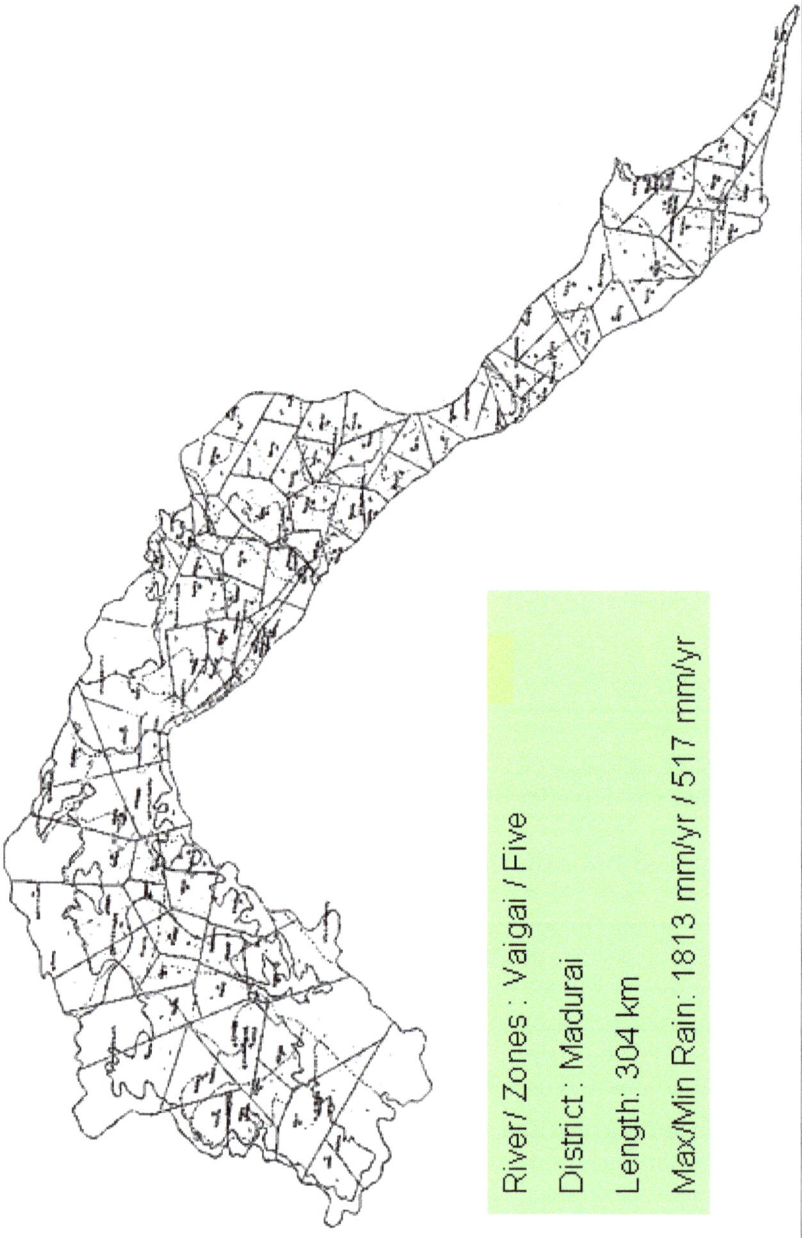

Figure 10.1: Theissen polygon for estimation of groundwater recharge.

contentious 999-year agreement between Kerala and Tamil Nadu. The groundwater elevation data from observation wells of the basin are taken from 120 wells data available at the government department and the data used for the study are tabulated in Annexure 1.

The western and northwestern parts of the basin receive heavy rainfall during the monsoons, with an average rainfall of 850mm over the basin. Hundreds of works have been carried out in Vaigai basin and most recent and important works are described here. The underlying geology of the basin is moderately complex; forming a combination of unconfined, confined, semi- confined and fully confined aquifers with significant spatial variation in productivity.

Earlier Works of the Vaigai System

Various organizations in India has carried out the study of water resources of Vaigai river basins and the differences between the present work and the earlier works are represented in Table 10.1. Some of them are described in detail in the following section.

Groundwater Division of *Public Works Department (PWD, 1976)* has conducted a groundwater balance study in Vaigai basin between Vaigai dam and Peranai. They have also conducted water quality studies on groundwater and shown that there is a general improvement in the quality of water in the few cross-sections of the basin. The rugged upland topography with the bedrock devoid of fissures and fractures are the reason for the insignificant addition to the groundwater in the upper reaches. The insignificant extraction of groundwater in the reaches I to III is due to the unfavorable sub-surface formations.

Table 10.1: Comparison with earlier studies on Vaigai

Sl.No.	*Earlier Studies*	*Current Study*
1.	Demands are for sub-basin wise only	Demands are for command area wise
2.	Groundwater modeling for estimation of aquifer parameters	Groundwater modeling for future pumping policy
3.	Water balance for entire basin	Water balance zone wise
4.	Yearly estimation of dynamic reserve of Groundwater	Monthly estimation of dynamic reserve of Groundwater

Institute for Water Studies (IWS, 1985) had carried out assessment, planning, development, and management of water resources of the Vaigai River basin. The MITSIM – MIT simulation model using IBM 370/155 computer was adopted for simulation studies. Historical data available for the years (1971-1982) were used for the simulation runs of the model. The water management strategies and institutional framework required for effective management of available water resources have been evolved for implementation. The remedial measures were suggested to control the erosion in the Vaigai basin area and siltation in the Peranai Regulator by active afforestation, development of orchards, plantation, gully plugging and by construction of check dam.

The *National Water Policy (NWP, 1987)* has laid great stress on the need for optimal water resources development. They have recommended many basins for study including Vaigai basin. Water and Power Consultancy Services (India) Ltd. (1996) carried out a Comprehensive study of Vaigai river basin. This encompasses assessment of surface and groundwater resources, their present level of use, the projected demand for the year 2015 AD for irrigation, municipal and industrial use, integrated operation of the reservoirs, macro planning of numerous tanks, conjunctive use etc., A simulation model to optimize the use of available resource of water in the Vaigai basin. Number of recommendations such as restoring Periyar operative FRL to 152 Ft., sharing of shortages in the ratio 2:5:4 for the zones III, IV and V, changes in gate operation schedule/cropping pattern, investigation of a storage site at identified location have been suggested. They have derived general conclusions on opening of gates for irrigation in June for the I crop of double crop areas of Cumbum and PMC and September 2nd fortnight for the II crop of double crop area of Cumbum and PMC, PMC main, PMC extension and for TMC main and extension.

Madras Institute of Development Studies (MIDS, 1997) has studied the condition and characteristics of groundwater utilized for irrigation in the basin. This study has examined the important features of groundwater irrigation such as statistical data, extent of water table decline, socio-economic and environmental implications, assessment of well irrigation investments and characterizes and functioning of groundwater markets in the basin.

World Bank (WB, 1999) developed a Decision Support System model and the Tools for the Holistic Analysis of Natural Network Information *(THANNI)* have been applied in Vaigai River Basin. The intricate network of thousands of traditional storage tanks needs structural and siltation rehabilitation. This DSS has two major components: an Information System to help organizing basic hydrologic, agricultural, urban and other information about the basin; and an optimization model that maximizes the benefits from water use subject to a variety of hydrological, economic, legal and policy constraints. Both the information system and the model have been made in a flexible manner to allow for consideration of different scenarios, data updates, model reformulation and further analysis.

The THANNI DSS used a variety of popular software, such as EXCEL spreadsheets and the GAMS optimization package to create an interactive graphical interface for ease of input and output. The optimisation model helps give greater insight into the physical and conceptual interconnections in the basin and the inter-sectoral and inter-regional tradeoffs involved in the consideration of various hardware (infrastructure) and software (policy) options for basin management. Currently, the model as formulated is a monthly model for the chosen planning year. The decision variables are the inter-sectoral water allocation, cropping pattern, flows in the system, reservoir operation, etc. The objectives are to maximize water use benefits subject to constraints on area available, meeting crop water requirements, meeting in-stream water use requirements, etc. The interactive model can be optimized for each scenario considered. THANNI gives decision-makers an additional tool for policy and scenario analysis and gives stakeholders a focal point for discussions. These should be augmented with expert opinion, views, policies, and other tools such as Geographic

Information Systems, and simulation, water quality, stochastic models, etc. as appropriate.

Institute for Water Studies (IWS, 1999) has carried out a micro level study of Water Resources plan for Vaigai river basin. They have done micro level study on assessment of annual surface water potential for75 per cent dependable year using SCS (Soil Conservation Services) method dividing the Vaigai basin into ten sub-basins. The annual groundwater assessment is made for all the ten sub-basins. The groundwater quality and suitability for drinking and irrigation purposes have been assessed by chemical analysis of samples from 84 observation wells established in Vaigai basin for ten years from 1984-1993. The general laws, regulations and institutional elements, which are applicable to this basin, are also drawn in this study.

Office of Chief Engineer, *State and Surface water resources data center (CE, 2000)* has identified Vaigai basin as first one for the value addition through the inputs from hydrology project. They developed a comprehensive and reliable hydrological data stations, covering all aspects of hydrology, meteorology, hydrometeorology, hydrogeology and hydrochemistry. They have also established the Hydrological Information System (HIS) to document the data collected through various discipline.

Center for Water Resource (CWR, 2000), Anna University carried out base line survey of irrigation commands including Lower Vaigai basin. The study envisages assessment of the adequacy of available water to service the planned command area. They have also conducted socio-economic survey and studied the environment and public health aspects of the basin.

The *present study* is aimed at assessing resources both surface water and groundwater and problems like flood, drought, waterlogging, sedimentation, environmental issues etc., including collection of necessary data; stating the developmental objectives in terms of goals, socio economic priorities etc., to identify technical alternatives and feasibility and developing an integrated system model to suggest policy guidelines. Overall, this is a venture on using advanced modeling tools and upto date data for developing a micro-system modeling as against macro-system modeling and to integrate with the macro-model.

Groundwater Modelling Strategy

With the increase in demand of water for competing uses, it is difficult to meet the entire demand from a single source and is a challenge to plan and optimally utilize the available water resources for proper management. The same situation exists for almost all irrigation systems in the world. It also exists in Vaigai basin as perceived from Figure 10.2.

Further the excessive development of groundwater in deltas of coastal areas may lead to salt water ingress in the freshwater aquifer and may contaminate it if unchecked. This may happen when sufficient surface water supplies do not reach the tail end of command areas and in the presence of overdrawal of groundwater.

Table 10.2 shows a comparison between the earlier and present works carried out in the water resources of Vaigai basin. Monthly available groundwater as dynamic

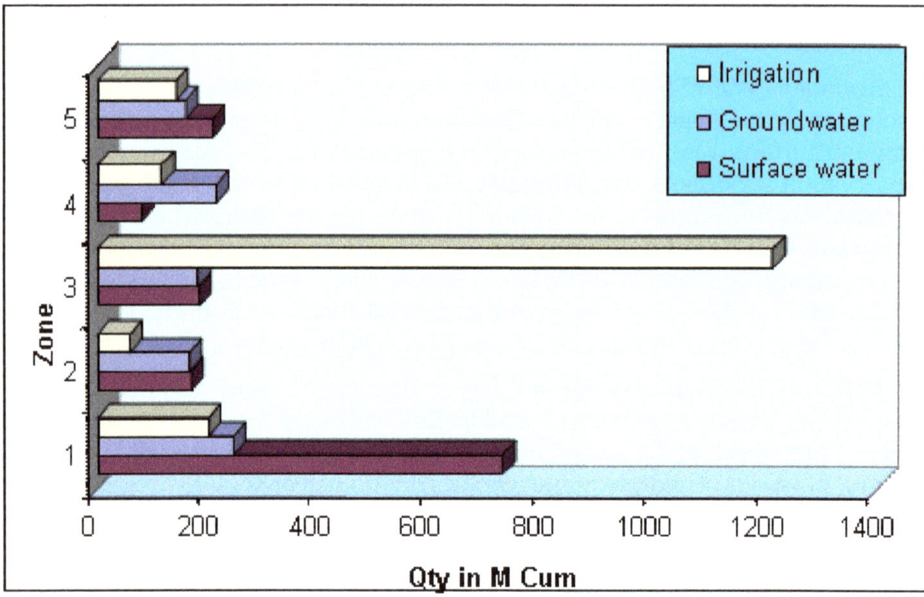

Figure 10.2: Surface and groundwater resources of Vaigai Basin.

reserve at each node in the basin is also estimated considering their ayacut as influencing area for respective nodes. An unusual rainfall on February 17 during 1989-90, which is the 75 per cent dependable year for groundwater, has caused a rise in groundwater in this month. The present work focuses on micro-level studies on the entire system by considering all outlets of the system from reservoir to the field channel. In addition the present work has employed recent computer softwares in all the segments of the study to achieve more accurate and quick results by using available database.

Table 10.2: Comparison of recharge estimation in Vaigai Basin

Sl.No.	*Year*	*Organization*	*Qty of G.W in Mcm in a Year*
1.	1999	By current study	972.82
2.	1997	Madras Institute of Development Studies	859.8
3.	1996	Water and Power Consultancy Services (India) Ltd.	1956
4.	1985	Groundwater Dept., PWD,Tamil Nadu	993

Groundwater Models

Modeling groundwater systems involve both water quantity (flow) and water quality considerations. Groundwater models incorporate mathematical representations of some or all of the following processes: movement of water and fluids through saturated or unsaturated porous media or fractured rock, transport of water soluble constituents, transformation of contaminants by chemical, biological and physical processes and heat transport and associated effects of temperature variations on groundwater flow and pollutant transport and fate of contaminants.

Classification of Groundwater Models

Groundwater models may be functionally characterized as (1) descriptive prediction models, (2) prescriptive management models and (3) parameter identification models. Most groundwater models are prediction models, which use input data characterizing the natural groundwater system and human-induced development to predict the hydraulic, and/or water quality conditions expected for specified management scenarios. The commonly used descriptive groundwater models simulate flow, with or without solute transport, in continuous porous media. Other more complex models simulate flow in fractured media, multiphase flow, or complex chemistry (geochemical models). This model provides descriptive predictions of the system response to a specified management plan. Prescriptive management models, on the other hand, are oriented towards determining optimal management plans for meeting specified objective. These models involve linking an optimization algorithm with a flow and/or solute transport model. Parameter identification or calibration models address the inverse problem of determining values for the parameters, which characterize the groundwater system.

There is a continuous development in the field of computer softwares applicable for groundwater studies. Some of them are *Flowpath, Flownet, Airflow, Prince, Aquifertest, Visual Groundwater, Visual Help* etc.

Groundwater Hydraulic Simulation by Visual Modflow

Model Development

Data Used

The groundwater annual recharge potential of Vaigai basin has been assessed using 28 yrs (1972-'99) of water level data. The observation well data collected from Groundwater Division are used. The data such as bore hole vertical section, map of observation wells, administrative boundary map and basin location map are used to develop the model. In order to construct the numerical model, geological borehole descriptions for pumping and observation wells, aquifer types and their properties are gathered to obtain an idea about the geological setting of the study area. Aquifer Properties: Most of aquifer area in the basin is found to be confined and semi-confined in which water table varies in undulating form depending on aquifer properties, areas of recharge, discharge and pumping from wells.

Geology

The main rock types are found to be Charnockites, Khondalites, Granites, limestones, and calcareous sand stones of archaean age. They are exposed in the plains of Melur taluk with acid intrusive pink granite and gneiss. Minor pegmatite and quartz veins are seen throughout.

Water Quality

The geochemicals such as calcium bicarbonate, calcium chloride, magnesium bicarbonate, magnesium chloride, sodium bicarbonate, and sodium chloride are present and thus a calcareous nature of soil is shown.

Generation of the Model

The model domain is created using the base map of Vaigai river basin is shown in Figure 10.3 with its well locations, solid boundaries etc. using world co-ordinates and model co-ordinates for better visualization. Model grids are smaller and more homogeneous than would be used in practice. The model is defined within the geo-referencing point as (225 km, 124km) fixing the boundary of the study area. The basin boundary is superimposed over the finite difference lattice. The generated model domain is shown in Figure 10.3.

Period of Simulation: Simulation of the model is run for pre-monsoon, monsoon and post-monsoon seasons of the river basin and water levels at each grid point are estimated. The simulation is performed with a groundwater modeling software Visual MODFLOW.

Validation of the Model

The properties of the aquifer such as hydraulic conductivity and storage co-efficient of the basin are plotted in regular contours. These provide an idea about the

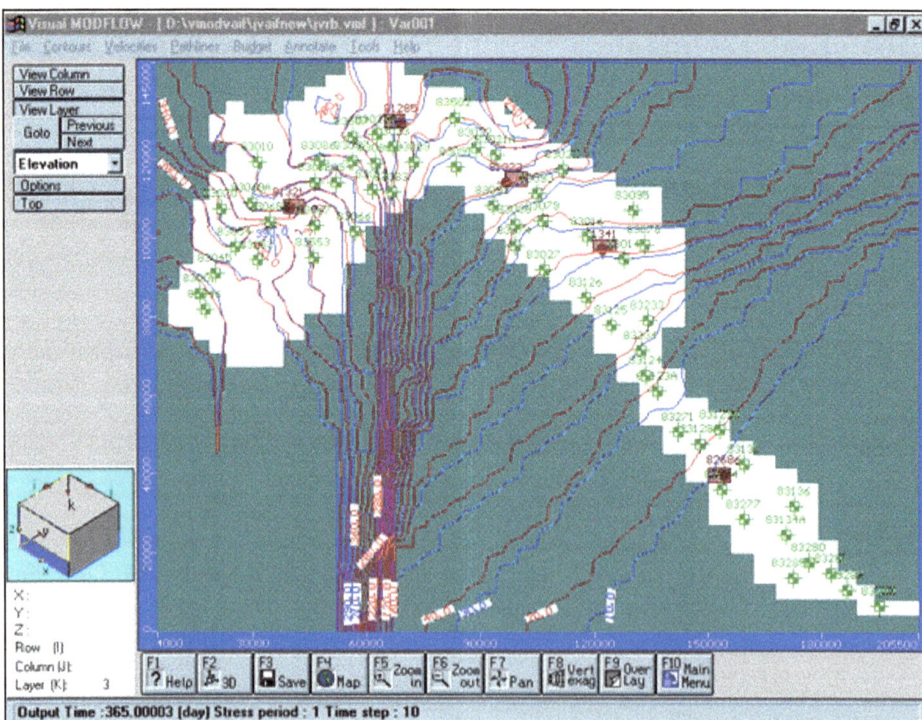

Figure 10.3: Vaigai Basin Model Domain with Wells and Elevation Contours

possible sources of water in the aquifers in the basin. The contours of head equipotential lines and water table contours are shown in Figures 10.4 and 10.5. Groundwater model, which is developed under static head, is validated with observed head Vs calculated head plots. The direction of velocity of flow in plain are obtained as output from Visual Modflow.

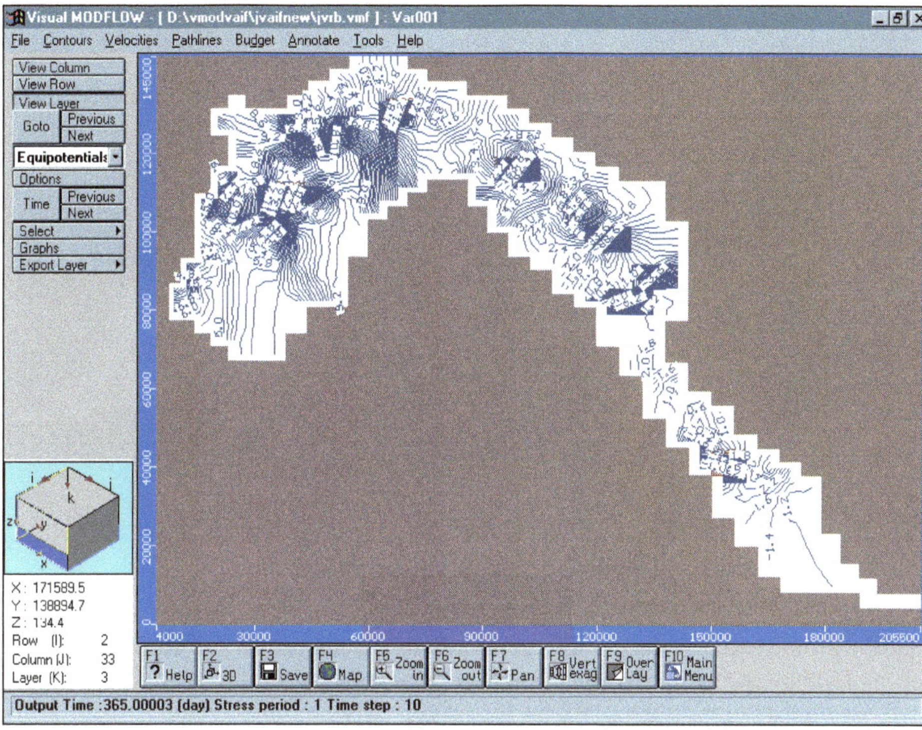

Figure 10.4: Head equipotential lines.

Results of Simulation

The computed groundwater level in the grid points has been taken and the average groundwater level for the basin is assessed. Flow associated with external stresses such as wells, areal-recharge, drains, and streams has been simulated.

The study is done for visualizing water table contours, direction of groundwater flow, the flow regimes on draw down contours, head equipotential, water table contours, recharge potential, layer thickness, flux between layers and head difference between layers of the aquifer. A color-coded plots of steady state and transient pathlines with time markers and velocity vectors are studied.

Conclustion

The most effective form of flow model results when the data are fitted to the model systematically. For this the simulation of the model is done for 63 wells and is compared with the observations made in the remaining 22 wells. Results indicated

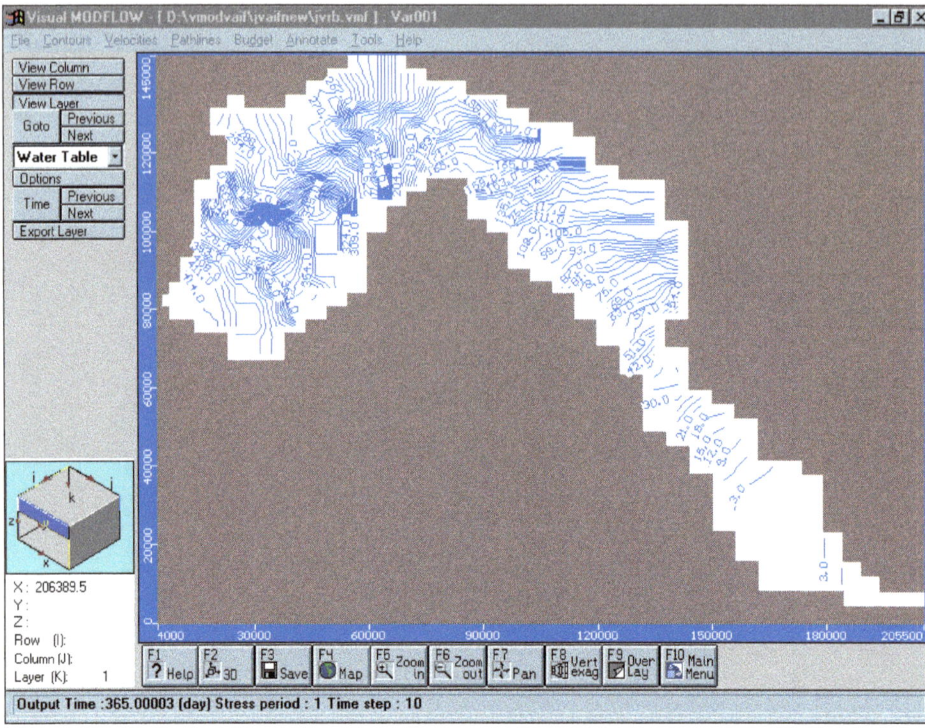

Figure 10.5: Water table contours.

that the groundwater levels would generally decline between 8 to 10metres within ten years, showing groundwater depletion in the basin. The calibrated model can be implemented as a guiding tool for the future groundwater management in the basin.

References

Environmental status report of vaigai river basin (prepared by the environmental cell of CE/P and D//PWD).

Flow data (daily readings) vol.8 in Palanichetty patty anicut(the last anicut in Cumbum valley form 6/74 to 6/94.

Flow data (daily readings) vol.9 in Palanichetty patty anicut (the last anicut in cumbum valley) from 7/94 to 12/99

Govt, of Tamil Nadu PWD, Periyar - Vaigai project Madurai. Comprehensive System Study of Vaigai River Basin (WAPCOS) Data Annexure Vol. 3/96.

India water resources management sector review, working paper 6, River Basin Modelling.

Samani, N., Kompani-Zare, M., Barry, D. A. (2004). MODFLOW Equipped with a New Method for the Accurate Simulation of Axisymmetric Flow. *Advances in Water Resources*, **27**: 31-45.

Senthilkumar, M., Elango, L. (2004). Three-dimensional Mathematical Model to Simulate Groundwater Flow in the Lower Palar River Basin, Southern India. *Hydrogeology Journal.* 12. 197-208.

State framework water resources plan, Annexture -10, Vaigai river basin, Institute of water studies, Taramani, Chennai.

ANNEXURE 1

Groundwater observation wells in Vaiqai basin

Sl.No.	*Available Well Data*	*Wells having Useful data*	*Sl.No.*	*Available Well Data*	*Wells having Useful data*
1.	83010 WK1	83010 WK1	35.	83069A WK1	83069A WK1
2.	83011 WK1	83011 WK1	36.	83086 WK1	83086 WK1
3.	83011A WK1		37.	83091 WK1	83091 WK1
4.	83012 WK1	83012 WK1	38.	83094B WK1	83094B WK1
5.	83014 WK1	83014 WK1	39.	83100A WK1	83100A WK1
6.	83014A WK1		40.	83117 WK1	83117 WK1
8.	83014C WK1		41.	83119 WK1	83119 WK1
9.	83015 WK1	83015 WK1	42.	83119A WK1	
10.	83016 WK1	83016 WK1	43.	83120B WK1	83120B WK1
11.	83016A WK1		44.	83121 WK1	83121 WK1
12.	83017 WK1	83017 WK1	45.	83121A WK1	
13.	83019 WK1		46.	83123A WK1	83123A WK1
14.	83019A WK1	83019A WK1	47.	83124B WK1	83124 WK1
15.	83020A WK1	83020A WK1	48.	83125 WK1	83125 WK1
16.	83021 A WK1	83021 A WK1	49.	83126 WK1	83126 WK1
17.	83022 WK1	83022 WK1	50.	83128AB WK1	83128A WK1
18.	83023 WK1	83023 WK1	51.	83129B WK1	83129B WK1
19.	83023A WK1		52.	83130AB WK1	83130A WK1
20.	83025 WK1	83025 WK1	53.	83130B WK1	
21.	83026 WK1		54.	83131A WK1	83131A WK1
22.	83026A WK1	83026A WK1	55.	83132A WK1	83132A WK1
23.	83027 WK1	83027 WK1	56.	83133A WK1	83133A WK1
24.	83028 WK1	83028 WK1	57.	83134A WK1	83134A WK1
25.	83045A-1 WK1		58.	83135 WK1	83135 WK1
26.	83045 WK1	83045 & 45B	59.	83136 WK1	
27.	83045B WK1		60.	83136A WK1	83136A WK1
28.	83046 WK1	83046 WK1	61.	83138 WK1	83138 WK1
29.	83053 WK1	83053 WK1	62.	83139B WK1	83139B WK1
30.	83054 WK1	83054 WK1	63.	83140 WK1	83140 WK1
31.	83066A-1 WK1	83066A-1 WK1	64.	83141A WK1	83141A WK1
32.	83068B-1 WK1		65.	83149A WK1	83149A WK1
33.	83068 WK1		66.	83151 WK1	83151 WK1
34.	83069 WK1		67.	83152 WK1	83152 WK1

Contd...

Annexure 1–*Contd...*

Sl.No.	*Available Well Data*	*Wells having Useful data*	*Sl.No.*	*Available Well Data*	*Wells having Useful data*
68.	83232 WK1	83232 WK1	90.	83284 WK1	83284 WK1
69.	83233 WK1	83233 WK1	91.	83285A WK1	83285A WK1
70.	83234 WK1	83234 WK1	92.	83286 WK1	83286 WK1
71.	83235B WK1	83235B WK1	93.	83287A WK1	83287A WK1
72.	83236 WK1	83236 WK1	94.	83288 WK1	83288 WK1
73.	83237 WK1	83237 WK1	95.	83289 WK1	83289 WK1
74.	83264 WK1	83264 WK1	96.	83290 WK1	83290 WK1
75.	83265 WK1	83265 WK1	97.	83290A WK1	
76.	83266 WK1	83266 WK1	98.	83291 WK1	83291 WK1
77.	83267 WK1	83267 WK1	99.	83292A WK1	83292A WK1
78.	83271 WK1	83271 WK1	100.	83296 WK1	83296 WK1
79.	83272A WK1	83272A WK1	101.	83297A WK1	83297A WK1
80.	83273 WK1	83273 WK1	102.	83298 WK1	83298 WK1
81.	83274 WK1	83274 WK1	103.	83299 WK1	83299 WK1
82.	83276 WK1	83276 WK1	104.	83300 WK1	83300 WK1
83.	83277 WK1	83277 WK1	105.	83301A WK1	83301A WK1
84.	83278 WK1	83278 WK1	106.	83302 WK1	83302 WK1
85.	83279A WK1	83279A WK1	107.	83303A WK1	83303A WK1
86.	83280 WK1	83280 WK1	108.	83304 WK1	83304 WK1
87.	83281A WK1	83281A WK1	109.	83305 WK1	83305 WK1
88.	83282 WK1	83282 WK1	110.	83554A-1 WK1	
89.	83283 WK1	83283 WK1	111.	83554 WK1	83554 WK1

11

Water Scenario and the River Linking Proposal: An Impact Analysis

Pallavee Tyagi & Anil K. Gupta

INTRODUCTION

"Flame of Life Burns in Water". Water forms an essential commodity of life at domestic and ecosystem levels. It plays a complex and multifaceted role in both human activities and natural systems. Rainfall, amount of utilizable water resources, and population pressures, are among the important factors that govern the level and pattern of water availability. India has four percent of world's freshwater resources. Twelve percent areas of the country receive an average rainfall of less than 610 million meters (mm) annually. Only eight percent areas receive more than 2500 mm of water. Cherrapunji in the eastern part of Meghalaya receives 11,000 mm of rainfall whereas the Western Rajasthan receives only 100 mm. Thus, this spatial and temporal variability in the water availability and both '**too little**' (*i.e.* Droughts) and '**too much**' (*i.e.* Floods) have negative effects on human well being (the global theme of UN-IDNDR, 1997). On the other hand, demand for water is expected to increase substantially in rural (agriculture) areas, industrial, power sector, and very importantly - for drinking. Water Stress Index, defined, as "the minimum per capita water required for basic households needs to maintain good health", expresses water availability in a region. A region is called as the "water stress region" if the yearly renewable freshwater availability is below 1,700 m^3/person. And the region is termed as "water scarcity region" where yearly availability falls below 1,000m^3 per capita (Falkenmark and Lindh, 1976). If no corrective measures are adopted, the current annual per capita

water availability of 1870 m^3 will be reduced to about 1340 m^3 by 2025, and according to TERI Director R. K. Pachauri, nearly 60 per cent of the country's population will be living in water stress conditions by then.

Problems of Uneven Distribution

The floods and droughts are very frequent problems in India due to monsoon variation. As per estimates 34 million hectares (m.ha.) are flood-prone of which 23 m.ha. are in the Ganga - Brahmaputra - Meghna basins in the states of UP, Bihar, West Bengal and Assam where the average annual total damage has been Rs. 972 crores (1953-1995). On the other hand, the drought prone area is estimated to be 57.12 m.ha. During year 2000-2001, eight states were affected involving 146.3 million people and 21.1 m.ha. of cultivated area.

Construction of dams for flood control is basically at conflict with their function of power generation or irrigation. According to Ramaswamy Iyer, former Secretary, Ministry of Water Resources, dams play only a modest role in flood moderation. Considerations for structural safety have sometimes necessitated the release of impounded waters causing 'man-made' floods. In Orissa, the frequency of floods in Mahanadi delta has actually increased after the Hirakund Dam. In the year 2000, heavy monsoons and the simultaneous release of waters from several dams led to severe flooding in West Bengal, affecting 21 million people in over 6000 villages. Controversy also surrounds the Tehri Dam because it has been built in a seismically unstable (zone 4) area despite warnings from the Geological Survey of India.

Water Development

The water requirements for all purposes are so large that a single developmental strategy cannot meet these on its own. Building of storage dams in India is inevitable for conservation of excess monsoon flows to meet the growing demand for water. Minor storages, watershed development through check-dams and use of groundwater are supplementary to major storages and not the alternatives. Since the possible storage sites are limited and spatial and temporal variation considerable, the National Commission for Integrated Water Resources Development Plan (NCIWRDP) has expressed the view that in a basin plan, there must be a place for a whole range of structures have localized benefits and applications, whereas large storages have a capability to transfer water over longer distances.

Inter-basin Transfer of Water: Historical Perspective

Long distance interbasin transfer of water is not a new concept and has been a practice in India for over five centuries. The Periyar Project, Parambikulum Aliyar Project, Kurnool - Cudappah Canal and The Telugu Ganga Project are good examples of interbasin water transfers executed in South India in 19th and 20th centuries. The *Periyar Project* is the most notable endeavor of the 19th century in transbasin diversion. A masonry gravity dam 47.28 m height was constructed across a gorge on the west flowing Periyar River. A 1740 m long tunnel with a discharging capacity of 40.75 m^3/s has been driven across the mountain barrier to convey the waters eastwards to the Vaigai basin. The project was commissioned in 1895 and provided irrigation to 0.58

lakh ha initially, which has since been extended to 0.81 lakh ha. There is also a power station of 140 MW capacities. *Parambikulum Aliyar Project* is a complex multi - basin multi - purpose project. Seven streams, five flowing towards west and two towards the east, have been dammed and their reservoirs interlinked by tunnels. The water is ultimately delivered to drought prone areas in Coimbatore district of Tamil Nadu and the Chittur area of Kerala State. The project has a command area of 1.62 lakh ha. and a total of 185 MW of power generation capacity. The *Kurnool - Cuddapah Canal* is 304 km long with a capacity of 84.9 m^3/s extending from Krishna to Pennar basin for irrigating an area of 0.53 lakh ha. The *Telugu Ganga Project* brings Krishna water from Srisailam reservoir through an open canal, first to Somasila, water is taken through a 45 km canal to Kandaleru and then to Poondi reservoir in Tamil Nadu through another 200 km long canal. By agreement among the Riparian states 12 TMC of water will be delivered to Tamil Nadu to augment water supply to Chennai city. The canal also irrigates 2.33 lakh ha. in Andhra Pradesh enroute.

Similarly in North India, inter sub – basin transfers in the Indus basin and Rajasthan canal project are some of the projects executed in the 19th and 20th centuries. *Inter Sub-basin Transfers in the Indus Basin*: A diversion dam, Pandoh, 140 km upstream of Pong Dam on Beas, enables diversion of water from Beas to Bhakra reservoir on Sutlej River and generates 165 MW of power on the way. The Beas – Sutlej link is 37.25 km long of which 25.45 km is tunnel through difficult rock formations. The Rajasthan Canal diverts water from the Himalayas to the deserts of Rajasthan. The project comprises of a huge multi purpose project constructed across the Beas River at Pong, a barrage at Harike and a Grand Canal system. These interlinking projects executed both in South and North India have been highly beneficial and have rarely been reported for noticeable environmental damages.

The United States, which is water rich and relatively scarcely populated, is transferring 45 BCM through interbasin transfers and has plans to add 376 BCM. Compared to this, India is transferring 10 BCM through existing schemes and has plans to add about 200 BCM. China has a scheme under implementation, which will transfer about 48 BCM. This indicates that India is already late in implementing the water transfer links.

Interlinking of Indian Rivers–Proposal

In 2002, Central Government of India announced to interlink India's major rivers by the year 2016. The plan, floated by the National Water Development Agency (NWDA), involves thirty links to transfer 173 billion cubic maters (BCM) of water through a 12,500 km. maze of canals. The aim is to ensure optimum utilization of freshwater by effecting transfers from areas of surplus to these of deficit. The total cost of the project is to be 5.6 lakh crore. The plan comprises two components *viz*: Himalayan Rivers Development and Peninsular Rivers Development (Table 11.1).

(*a*) Himalayan Rivers Development

Himalayan Rivers' Development envisages construction of storage reservoirs on the principal tributaries of the Ganga and the Brahmaputra in India, Nepal and Bhutan, along with interlinking canal systems to transfer surplus flows of the eastern

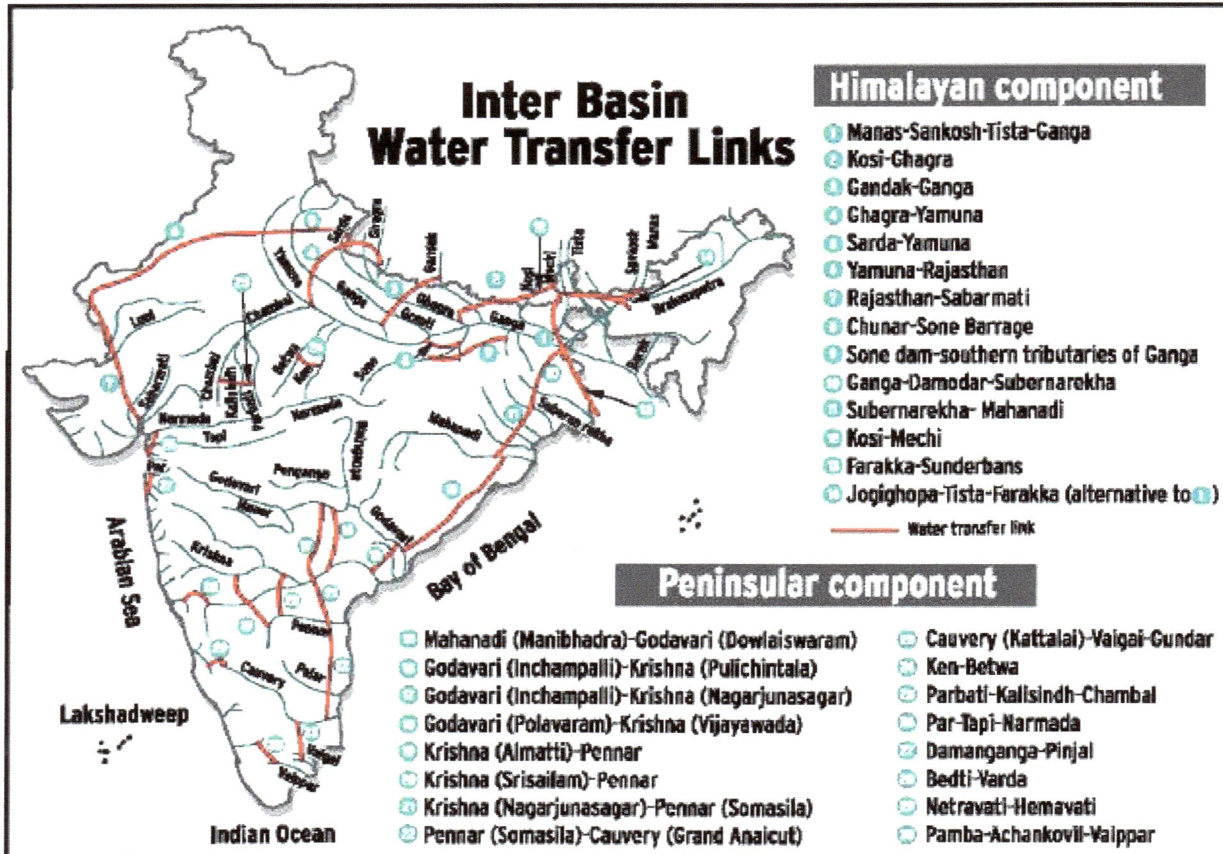

Source: National Water Development Agency

Figure 11.1: River Linking Proposal of India.

tributaries of the Ganga to the west, apart from linking of the main Brahmaputra and its tributaries with the Ganga and Ganga with Mahanadi.

Table 11.1: Proposed inter-basin water transfer links

	Himalayan Component		*Peninsular Component*
1.	Brahmaputra – Ganga (MSTG)	16.	Godavari (Inchampalli low dam) – Krishna (Nagarjuna Sagar tail pond)
2.	Kosi – Ghagra	17.	Godavari (Inchampalli) – Krishna (Nagarjuna Sagar)
3.	Gandak – Ghagra	18.	Godavari (Polavaram)-Krishna (Vijaywadw)
4.	Ghagra – Yamuna	19.	Krishna (Almatti)-Pennar
5.	Sarda – Yamuna	20.	Krishna (Srisailam)-Pennar
6.	Yamuna – Rajasthan	21.	Krishna (Nagarjunasagar)-Pennar (Somsila)
7.	Rajasthan – Sabarmati	22.	Pennar (Somsila)-Cauvery (Grand anicut)
8.	Chunar-Sone barrage	23.	Cauvery (Kattalai)-Vajgai-Gundar
9.	Sone dam – Southern tributaries of Ganga	24.	Ken-Betwa
10.	Ganga – Damodar – Subernarekha	25.	Parbati-Kalisindh-Chambal
11.	Subernarekha – Mahanadi	26.	Par-Tapi-Narmada
12.	Kosi – Mechi	27.	Damanganga-Pinjal
13.	Farakka – Sunderbans	28.	Bedti-Varda
14.	Brahmaputra (ALT)-Ganga (JTF)	29.	Netravati-Hemavati
15.	Mahanadi (Manibhadra)-Godavari (Dowlaiswaram)	30.	Pamba-Achankovil-Vaippar

(*b*) Peninsular Rivers' Development

This component is divided into four major parts:

1. Interlinking of the Mahanadi - Godavari - Krishna - Cauvari Rivers and building storages at potential sites in these basins. This is the major interlinking of the river systems where surpluses from the Mahanadi and the Godavari are intended to be transferred to needy areas further south.
2. Interlinking of West flowing rivers, north of Mumbai and south of Tapi. The scheme provides for taking a water supply canal to metropolitan areas of Mumbai. Gujarat State would get benefit of irrigation.
3. Interlinking of Ken - Chambal Rivers, the scheme provides for a water grid for Madhya Pradesh and Uttar Pradesh through an interlinking canal backed by storages.
4. Diversion of other west flowing rivers to meet the requirements of Kerala as also for transfer of some waters to east to meet the needs of draught affected areas.

Projected Benefits

The extent of work involved in the entire project is shown in Table 11.2.

Table 11.2: Quantum of work involved in the project

Works Involved	*Quantity*
Total earthwork envisaged	55,890 m.cu.m.
Total concrete work envisaged	3,507 m.cu.m.
Total steel Utilized	30 m. tones
Total unskilled labour days	9,953 crore days
Total semi-skilled labour days	360 m. days
Total skilled labour days	1.25 m. days

Source: Dahasahasra *et al.*, Journal of Indian Water Works Association, 2003.

The river-linking project will help to add 34,000 MW generation of hydroelectricity. Advantages of the project are flood and drought control, improved agriculture, alternative transport, higher GDP growth (4 per cent), decentralization of cities, social benefits, boost in fisheries and tourism. Major benefits are shown in Table 11.3.

Table 11.3: Major benefits of the project

Benefits of Works	*Quantity*
Total surplus crop yield	165 million tons
Total Arid regions covered	56 million hectares
Total drinking water available	30,000 million cum
Total hydropower generated	57227 MW
Total extra roads laid	38,245 kilometers
Total freshwater fishing area	1.2 million hectares

Source: Dahasahasra *et al.*, Journal of Indian Water Works Association, 2003.

Major environmental profits from the project are improved public health due to water supply, sanitation, reduced pressure on stressed supply system, reduced use of local water for irrigation, reduced pressure on over drafted groundwater and reduced pollution in some of the rivers.

Problems Expected Due to Project Implementation

Critics appear that the interlinking plan is economically prohibitive, fraught with uncertainties and has the potential for disastrous and irreversible after-effects. Inter basin transfers are complex issues which have profound environmental, social, economic and political implications. Two such prior proposals, the Ganga - Cauvery link by Mr. K. L. Rao (1972) who was the then Minister for Power and Irrigation and the Garland Canal Plan by Capt. Dastur (1974) were rejected by the Central Water Commission on grounds of being techno-economically unviable and offering no flood-control benefits. Further, the National Commission on Integrated Water Resources Development (NCIWRD) Plan in its September 1999 Report, did not envisage the need for massive inter basin transfers. It was argued that the scarcities could be redressed through full development and efficient utilization of intra basin resources. As suggested by several economists, environmentalists and social activists, the following issues serious considerations by those supporting the mammoth enterprise.

Environmental Implications

By its own admission, the NWDA's proposal involves the construction of certain large dams. In its report "Dams and Development: A New Framework for Decision-Making (2000)", the World Commission on Dams has mentioned that 60 per cent of the world's rivers have been affected more adversely than positively by dams and diversions. In many cases, these have led to irreversible loss of species and to ecosystems. The review also reported the displacement of 40-80 million people globally by dams.

Dams and Flood Control

Construction of dams for flood control is basically at conflict with their function of power generation or irrigation.

Agriculture Practices

The greater availability of water combined with strong state support could prompt a preference for growing water-intensive cash crops, such as sugarcane. In Maharashtra, for instance, sugarcane is actually grown in drought prone areas, which makes continuous irrigation necessary at the rate of 2800-3000mm/year. Nearly 60 per cent of water from irrigation projects is used to water 500,000 hectares of sugarcane growing land, which leaves little or no water for non-sugarcane farmers (Shrivastava *et al.*, 2011). Yet another grave problem is that of waterlogging and salinity.

Economic Issues

Apart from the estimated cost of Rs. 5.6 lakh crore, recurring expenditures would be incurred in maintenance of dams, de-silting reservoirs, re-lining canals and creating artificial drainages where needed. Other expenses would include restoration of forests, sanctuaries, archaeological and heritage sites. Time and cost overruns typically associated with all major projects is also a distinct possibility. The NCIWRD has estimated an amount of Rs. 70,000 crore in the tenth plan and Rs. 110,000 crore in the eleventh plan for completion of spillover of ongoing projects (Iyer, 2001).

Epilogue

Prior experiences have taught us to study the basic aspects of each river basin, including its catchment area treatment, command area development, the affected population, and impacts of reservoir and canal system on farmers, and fisheries, and public health, as a prerequisite for decisions about any major project. Increase in the scope of power generation increase from this project also needs to be examined realistically. Doubts about the flood and drought mitigation outcomes of this project are also prevailing. The impacts of climatic and agro-ecological alterations in different regions as a result of the inter-basin transfer, socio-cultural constraints, traditional wisdom issues, call for a systematic study for basin-wise ecological implications must have been done before conceiving such strategic proposals. Through

Environmental Impact Assessments, Cost-benefit (multi-criteria) analysis, quantitative assessments of non-quantifiable considerations, and based on these, rigorous investment appraisals, will need to be undertaken. Even more serious is the funding problem that has to be assessed in view of other short-term and long-term developmental requirements. (Acknowledgement: A range of national, international and web-based resources have been studies in making of this article, and authors extend acknowledgement with responsibility of useful academic interpretation).

References

Falkenmark and Lindh 1976, quoted in UNEP/WMO. "Climate Change 2001: Working Group II: Impacts, Adaptation and Vulnerability." UNEP. Retrieved 3 February 2009.

Iyer, R 2001. Water: Charting a Course for the Future. II. Economic and Political Weekly, 1235-1245.

Shrivastava, A.K., A.K. Srivastava and S. Solomon, 2011. Sustaining sugarcane productivity under depleting water resources. *Current Science,* **101(6)**: 748-754.

World Commission on Dams 2000. Dams and Development: A New Framework for Decision-making: The Report of the World Commission on Dams, Earthscan.

12

Water Use in Drinking: Analysis at Urban Local Body Level

Jharna Pathak

INTRODUCTION

The right to adequate drinking water is considered to be the most fundamental of human rights.

Since water is essential to life, the first priority is that it must be made available to one and all.

Availability of water is highly uneven in both space and times but have competing uses (Pathak, 2006). Realising the importance and scarcity attached to the freshwater, it has to be treated as an essential environment for sustaining all life forms. Gleick, 1999 argues that basic needs" go beyond what we need to drink or ingest through our food for daily survival (about 5 litres per person per day- it includes the need for water to maintain a basic standard of personal and domestic hygiene sufficient to maintain health. It is not sufficient merely to have access to water in adequate quantities - the water also needs to be of adequate quality to maintain health and it must be free of harmful biological and chemical contamination. Water gathered from unprotected sources often does not meet these criteria and places the users at risk. Often water, which is of a sufficiently high quality at the point of collection, is contaminated before it is used, as it has to be carried and stored before use.

Planning and implementation of water resource projects involve a number of socio-economic aspects and issues such as environmental sustainability, appropriate resettlement and rehabilitation of project-affected people and livestock, public health concerns of water impoundment and so on there is a need for common approaches and guidelines on these matters. Moreover, certain problems and weaknesses have affected a large number of water resources projects all over the country. There have been substantial time and cost overruns on projects. Moreover, growth process and the expansion of economic activities inevitably lead to increasing demands for water for diverse purposes: domestic, industrial, agricultural and so on. So far, the major consumptive use of water has been for irrigation. Complex issues of equity and social justice in regard to water distribution between uses and within users are required to be addressed. The development, and overexploitation of groundwater resources in certain parts of the country more importantly in the state like Gujarat and Rajasthan have raised the concern and need for judicious and scientific resource management and conservation. In National Water Policy (NWP 1987 and 2002), Government of India addressed these concerns on the basis of common policies and strategies. This policy is of an advisory nature and is intended to be used by national or regional authorities as a basis for the development of drinking-water standards and regulations appropriate to their own socio-economic and exposure situation. The *Guidelines* clearly recognizes that irrigation and multipurpose projects should invariably include a drinking water component, wherever there is no alternative source of drinking water.

Since 80's all large-scale irrigation projects in Gujarat have been termed as multipurpose Irrigation projects with prioritizing drinking water requirement. However, urban local government across India has failed to meet the challenges of increase in water requirements. The urban local government of towns in Gujarat is no better. Local government mainly Panchayats for villages, municipalities in towns (consisting of population)and Municipal Corporation (with the population) are charged with the duty to supply BWR (70 lpcd for towns without drainage and 40 lpcds for Panchayats) water to its citizens.

The problem of drinking water is often blamed on the natural water scarcity in the regions and low rainfall. As a solution, water is brought from outside in tanks or trains or else is transferred from water abundant reservoir to water scarce reservoir. In this paper, I discuss how at the time of water scarcity, this drinking water demand exerts pressure on water available in the reservoir leaving little for irrigation use. In the backdrop of challenges created by ever-increasing population, depleting groundwater table and fixed basic water requirement norms, this paper discusses status of water distribution by municipalities. A deeper understanding of the performance of water use in drinking water sector will throw light on Basic Water Requirement (BWR) policy and assess the absorption capacity of water use in this sector.

The paper brings out the point that management plans for water supply heavily rely on engineering and structural solutions to meet ever-growing demand for water but pay little attention to non-structural alternatives. It brings out that the prevailing view among administration that water as an essential and a basic human need has unfortunately created strong pressures on government-owned utilities to subsidise water tariffs. This paper discusses that myopic policies of the government aggravate conflicts of water demand between planned users and actual users. This paper also bring out the extent of gap that exists between water charges and costs of supplying water to its users, which is expected to throw light on financial sustainability of local bodies.

This paper is based on water allocation from Dharoi reservoir project, located in Kheralu, Mehasana District of Gujarat State. This reservoir project suffers from problems of water scarcity, multiple claims on the use of water from irrigation and drinking sector and depleting groundwater level. Besides analysing water allocation from the reservoir between farm sector and drinking water sector, with the help of case study, this paper also discusses contribution and performance of water use in municipality. For the purpose of this paper, we limit ourselves to five municipalities, which receive water from the reservoir.

This paper proceeds as follows. Section II discusses characteristics of Dharoi Reservoir Project and the methods of water distribution among lean and peak seasons. Section III shows status of water availability in various municipalities receiveing water from the reservoir. The case study of Palanpur municipality is examined in section IV. Section V puts forth some policy issues in light of the evidence obtained.

Characteristic of Dharoi Reservoir Project

Dharoi Reservoir Project supplies nearly 24 percent of total water in the reservoir to meet the demand for drinking use (in the year 2001-2002) and the rest is diverted to the farm sector. To understand water supply from the reservoir, it is imperative to take into account total water available in the reservoir and water supplied from the reservoir to various sectors. Scholars have noted that allocation of water resources is determined not only by annual variability, but also by seasonal and monthly variability. In this paper, the analysis of total water supply from the reservoir is based on the information about the volume of water inflow and outflow of the reservoir, difference in storage and so on which are obtained from the Irrigation Department of Dharoi Reservoir Project.

At the reservoir level, in a given period, water inflow is the algebraic sum of difference in opening balance and closing balance of storage and water supplied for various uses. If the opening balance is higher than the closing balance, water available can be calculated by subtracting dead storage and closing balance of the current period from the net outflow. This inflow data is very important for studying water availability to various uses in the dynamic situation of the reservoir. As the information from reservoir consists of water stored in the reservoir on a monthly basis, computing total water availability to various uses over an entire year can be a reasonable first indicator for studying water situation in a reservoir. An attempt has been made to calculate yearly water availability from a fortnightly available data on water balance for the past sixteen years starting from 1986-87 up to 2001-2002. This when compared with Full Reservoir Capacity level (FRL) would throw light on peak and off peak periods. Details of water availability in the reservoir and water supplied to various uses are given in Table 12.1.

Water Supplied from the Reservoir during Peak and Lean Years

It is evident from Table 12.1, that in years 1988-89, 1990-91, 1991-92, 1992-93, 1993-94 and 1994-95, water availability in the reservoir was at its peak and the reservoir was filled to more than 80 per cent of its FRL. These years are termed as peak years. The years during which the reservoir was filled to forty per cent of its capacity are termed as lean years in this paper. These years are 1986-87, 1987-88, 1989-90, 1996-97, 1998-99, 1999-2000, 2000-01 and 2001-02. However, the years during which the reservoir is filled between forty per cent of its capacity and eighty per cent of the capacity is termed as normal year. These years are 1989-90, 1995-96 and 1997-98.

Yearly analysis shows that in normal years and water deficient year of 1995-96 and thereafter, water supply to various users was higher than water available in the reservoir. This resulted from an over-exploitation of storage. Barring two initial years of the project along with the year 1997-98 and 2001-02, in lean and normal years, more water was supplied for drinking water than that to the agricultural sector (Table 12.1). As the water availability was quite low, this increase in demand was met by utilising water stored in the reservoir during peak periods. Thus, water stored during peak years act as a buffer against increasing stress during water scarce years. This negates the existing argument that irrigation department should shun away from the practice of storing water in the reservoir in order to reduce evaporation losses in the semi-arid regions. Similarly in the year 1999-2000, when water in the reservoir was only two per cent of designed live storage, cent percent water from the reservoir was allocated to meet drinking water demand. Table 12.1 reveals a clear trade-off between agricultural and drinking water use during the lean period. This means that though currently water use for drinking purpose occupies only a small proportion of demand for water from the reservoir in comparison to farm use, but by prioritizing drinking water use (GOI, 1987 and 2002), groundwater overdraft and increase in urbanisation has the potential to exert pressure on reservoir water to be diverted to drinking water use.

Water Distribution from the Reservoir

As in many reservoirs of the country, water from Dharoi Reservoir was planned to meet drinking water needs of Ahmedabad and Gandhinagar cities in addition to supplying water for irrigation. However, the actual allocation of water from the scheme for drinking water sector was 730 million litres a day. In doing this, water had to travel the distance of more than 90 kilometers and the fallout of such a water distribution was losses due to theft and evaporation increased exponentially. Moreover, there were other areas like Mehasana and Patan districts, which were facing problems of groundwater overdraft. To resolve this challenge, Government of Gujarat has launched Regional Water Supply and Sanitation Programme (RWSS) to facilitate Gujarat Water Supply and Sanitation Board (GWSSB) to receive bulk of the water from the reservoir. The latter then distributes water to various municipalities and Gram Panchayats[1] facing the problem of inadequate water supply or severe problems of fluorosis. Following National Water policy (1987), under this programme, 70 litres per person per day (lpcd) of water was planned to be supplied to users from the reservoir in urban areas and 40 lpcd in rural areas. Thus, in the year 2000,

Table 12.1: Details of water supplied in various uses from Dharoi reservoir

Year	*Percentage of reservoir Filled (per cent to FRL)*	*Closing Balance (in mcm)*	*Exploiting Dead Storage (in mcm)*	*Total Inflow (in mcm)*	*Total Losses (in mcm)*	*Total Water Supply to Various Uses (in mcm)*	*Agricultural Use*	*Drinking Water Use*
							(Per cent to total water use)	
Lean period								
1986-1987	17.0	95	-35	206	67	209	100.0	0.0
1987-1988	1.0	75	-55	125	51	95	100.0	0.0
1996-1997	37.8	204	74	654	85	385	45.0	55.0
1998-1999	33.2	59	-29	389	75	362	39.6	60.4
1999-2000	1.7	64	-24	91	39	47	0.0	100.0
2000-2001	8.4	175	87	266	68	88	0.0	100.0
2001-2002	39.6	126	38	373	79	342	76.4	23.6
Peak period								
1988-1989	82.6	212	82	862	108	619	44.2	55.8
1990-1991	100.0	424	294	1005	142	535	60.8	39.2
1991-1992	99.7	178	48	817	159	905	71.7	28.3
1992-1993	100.0	327	197	1040	144	747	66.0	34.0
1993-1994	95.5	164	34	666	152	678	62.7	37.3
1994-1995	100.0	350	220	1060	171	702	64.5	35.5
Normal period								
1989-1990	49.7	96	-34	456	101	471	38.4	61.6
1995-1996	55.5	120	-10	426	125	531	43.3	56.7
1997-1998	63.8	107	-23	509	107	598	60.1	39.9

Source: Figures of closing balance, losses, total water use, inflow, and water use in various uses are collected from Irrigation Division Office, Visnagar.

Table 12.2: Details of villages covered and cost per capita of supplying drinking under regional water supply and sanitation scheme

Sl.No.	*Name of Schemes under Regional Water and Sanitation Programme (RWSS)*	*Number of Villages and Towns Covered*		*Population 1991 (in thousand)*			*Decadal Cost per capita(Rs per capita for the year 1998-2008)*	*Villages affected by Fluoride (in numbers)*
		Villages	*Towns*	*Villages*	*Towns*	*Total*		
1.	Dharoi RWSS	378	5	699 (68.33)	206 (89.57)	905 (66.94)	228,786	92
2.	Chansma Pilot project -1	34		71 (6.94)		71 (5.25)	239,422	34
3.	Chansma Pilot project -2	77	2	123 (12.02)	24 (10.43)	147 (10.87)	159,245	29
4.	Danta, Vadgam RWSS	107		130 (12.71)		130 (9.62)	165,365	107
5.	Palanpur city Scheme		1		1 (0.43)	1 (0.07)	416,902	10
	Total schemes under RWSS	**596**	**8**	1023 [75.66]	230 [17.01]	1352	229,496	416

Source: GWSSB division office, Visnagar (2003).

Note: Figures in parentheses indicate percentage to total and figures in [ ] indicate percentage to total population.

Government decided to stop meeting the drinking water needs of Ahmedabad and Gandhinagar cities from the Dharoi Project and diverted the water thus saved to fluoride-affected villages of Mehasana and Patan districts, which were dependent on tube wells for their drinking water needs.

The agenda of RWSS programme is to deal with problems of water supply by constructing water networks and provide assistance to local bodies in order to bring about user participation in water management. This necessitated the need to construct water networks in 596 villages and 8 urban towns with the total population of 1,352 thousand (Table 12.2). Currently the construction work on water network is complete in five urban towns namely Kheralu, Siddhpur, Unjha, Vadnagar and Palanpur.

Status of Water Availability

As work on water networks were complete in five municipalities namely Kheralu, Siddhpur, Vadnagar, Unjha and Palanpur, an attempt is made to analyse status of water availability in these municipalities with the help of case study. Table 12.3 showed the basic characteristics of five municipalities.

Table 12.3: Basic characteristics of study municipalities (2003-04).

Characteristics	*Municipalities*					
	Kheralu	*Siddhpur*	*Vadnagar*	*Unjha*	*Palanpur*	*Total*
Population (nos.)	29630	50770	21041	53000	125,000	283,671
Total households (nos.)	5000	13500	7000	14000	13,849	51,849
Area (sq kms.)	37.0	11.5	17.2	15.0	10.0	91.0
Total connections (nos.)	3,600	11,578	4,350	9,900	13,978	43,406
Percentages of residential tap connections to total connections (per cent)	94.4	98.1	97.7	98.5	93.8	96.5
Percentages of commercial tap connection to total (per cent)	2.8	1.9	2.3	1.5	5.3	3.0
Percentages of stand posts for Economically Weaker Sections (EWS) to total (per cent)	2.8	–	–	–	0.9	0.5
Area covered with access to piped water (per cent)	100.0	100.0	100.0	85.0	80.0	95.0

Source: Field survey (2003-04).

Among all municipalities, population of Palanpur municipality was higher compared to other municipalities. Barring Palanpur where pipe network is under construction and Unjha where groundwater is still a reliable water source, access to piped water is better in other municipalities. With nearly 97 percent of connection is residential connection (Table 12.3), improved access to piped water shows that municipalities are able to provide reliable and safe drinking water.

---

1. Municipalities in towns and Panchayats in rural areas were made responsible for distributing drinking water to their users.

Table 12.4 shows growth rate of population exert pressure on demand for available water.

Table 12.4: Decadal growth rate (in percentage) of population of study municipalities

Municipality	*Years*			
	1961-71	*1971-81*	*1981-91*	*1991-2001*
Siddhpur	19.7	28.2	-2.3	8.3
Unjha	33.4	38.8	35.2	3.9
Kheralu	33.2	27.7	3.0	62.2
Vadnagar	19.8	14.2	14.0	-16.4
Palanpur		66.3	31.7	55.0
Total Study Municipalities		38.5	18.4	25.6

Source: Field survey (2003-04).

Following GOI (1987), Municipalities supply BWR of 70 lpcd to each of its users. When quantity of water actually supplied by all municipalities was compared with the BWR norms it was clear that there has always been a shortfall. It was observed from data that since 1997, water supply had been declining over the years and it was nearly 45 lpcd in the year 2001 (Table 12.7). As groundwater remained the largest source for water use till the year 1999 (Table 12.5), shortage of water supply was largely due to inadequate water availability from groundwater source.

Table 12.5: Percentage of total water withdrawn from tubewells

Year	*Municipalities*					
	Kheralu	*Siddhpur*	*Vadnagar*	*Unjha*	*Palanpur*	*Total**
1997	100 (1,456)	100 (3,156)			100 (6,278)	100 (10,890)
1998	100 (1,424)	100 (3,156)			100 (5,987)	100 (10,566)
1999	100 (1,326)	100 (3,156)	100 (1,049)		100 (5,697)	100 (11,228)
2000	100 (1743)	0 (3353)	100 (1279)		100 (4600)	70 (10,976)
2001	100 (1,979)	0 (3,562)	100 (1,584)		29 (5,650)	41 (12,775)
2002	59 (2,222)	0 (3,695)	0 (1,883)	65 (4,681)	27 (5,600)	32 (18,082)
2003	0 (2,222)	0 (3,789)	0 (2,104)	65 (4,831)	25 (6,000)	24 (18,947)

The total percentage is the percentage of water withdrawn from groundwater. The rest is supplied by Dharoi reservoir.

Figures in parentheses indicate total water withdrawn by municipalities from groundwater and surface water (in '000 litres)

Source: Field survey (2003-04).

Table 12.6: Rate of growth of water withdrawn from tubewell

Year	*Growth Rate of Water Withdrawn from Tubewell*					
	Kheralu	*Siddhpur*	*Vadnagar*	*Unjha*	*Palanpur*	*Total**
1998	-2.24				-4.6	-3.00
1999	-6.88				-4.8	6.22
2000	31.49				-19.3	-31.79
2001	13.50				-64.1	-31.57
2002	-34.28				-9.1	11.04
2003					0.0	-20.22

Source: Field survey (2003-04).

As a result of declining water supply from groundwater, and launching of RWSS, municipalities that were till then dependent on groundwater for their drinking water requirement, started getting connected through a water network from the Dharoi basin in a phase wise manner. Siddhpur municipality became the first to avail of drinking water facility from this project. In the year 2000, it was hundred per cent dependent on canal water. Soon Palanpur, Kheralu, Unjha and Vadnagar joined the former in drawing water from GWSSB. By the year 2002, dependence of municipalities on surface water increased significantly. This situation is evident in Table 12.6, which shows that the proportion of water from groundwater has reduced from a high of 70 per cent in the year 2000 to 24 per cent in the year 2003. To improve municipal water supplies, the work on RWSS project was taken up in the year 2000. The impact of this work became visible due to an increase in water supply per capita for municipalities (Table 12.7). Thus during this year, water withdrawn from surface water increased from 30 percent in the year 2000 to 76 percent in the year 2003 (Table 12.5). Municipality wise study (Table 12.7) showed that Kheralu, Siddhpur and Vadnagar became fully dependent upon surface water, which was made available under RWSS project.

The status of importance of water accessibility against the norms of BWR as prescribed by NWP was studied by comparing water supplied to municipalities and BWR norms. Prior to RWSS (before 2002), demand per head was pegged at around 55 lpcd. Table 12.7 reveals that in the year 2003, average water supplied by all the municipalities under study was 73 lpcd per capita which was higher than the BWR norm. As water network in Palanpur municipality was incomplete (receives 48 lpcd of water only), average water supplied in all municipalities other than Palanpur shows that water supplied is significantly higher than prescribed norms of BWR (80 lpcd in 2003).

Municipality wise analysis shows that as this project was implemented in a phased manner some municipal areas, which got connected earlier, received a lion's share of the water while some were left high and dry. Table 12.7 showed that most municipalities were able to supply more water to their users than the norms specified by the NWP (75 lpcd in Kheralu, 84 lpcd in Vadnagar and 91 lpcd in Unjha). This showed that norms of BWR and improvement in water supply over the years have encouraged locals to demand more water. Moreover, the concept of specifying BWR norms to treat water as a basic good provided no incentive to use water efficiently.

Table 12.7: Details of quantity of water demanded and withdrawn from GWSSB and water supplied by municipalities to the users

(Quantity in litres per capita per day)

Year	*Water Supplied by Municipalities to its Users*						
	Kheralu	*Siddhpur*	*Vadnagar*	*Unjha*	*Palanpur*	*All (Average)*	*All Municipalities Except Palanpur (Average)*
1997	60	60			51	57	60
1998	56	59			49	55	58
1999	50	59			46	52	55
2000	63	62	57		37	55	61
2001	70	66	71		46	63	69
2002	77	68	56	89	45	67	73
2003	75	69	84	91	48	73	80

Source: Field survey (2003-04).

Current practice of municipalities of supplying water more than the specified norms would conflict with situation when water networks get completed. For example, completion of works of water network in Palanpur municipality would exert pressure on existing users. This exposes the limitation of BWR approach. Further, this norm was set uniformly without considering distinctly the commercial and residential uses of water in a municipality. Consequently, per capita water supply in the year 2003 was greater in smaller municipalities (Unjha and Vadnagar) than in the larger ones (Siddhpur) (Table 127). Thus, in smaller municipalities with few commercial establishments, more water was made available for residential use. These points to the fact that too much dependence was placed on supply side approach in order to provide water as per the BWR norms. As a result, Municipalities failed to provide incentive to users to save water. Such myopic planning was likely to encourage wasteful use of water.

The Specific Case of Palanpur Municipality

The above section examines water use by various municipalities. Given the fact that local bodies incur expenditure on supplying drinking water to their users and earn income in form of water charges along with other income sources, data on revenue and expenditure incurred in supplying water by municipalities were sought from other municipalities also. However, Palanpur municipality was the only one municipality, which provided data on revenue and expenditure incurred in supplying water to its people along with the information on total water supplied by municipalities. Therefore, the analysis on the extent of gap between the revenue earned by municipalities in supplying water and costs of supplying water got restricted to this municipality only. These costs and income are relevant in assessing the efficiency of existing supply mechanisms.

Some of the basic characteristics of Palanpur Municipality are

All Households Connected but the Revenue still Abysmal

From Table 12.8, it was visible that almost 100 per cent of the households have water connections. Moreover, nearly 99 percent of the population was covered with piped network (Table 12.8). Thus, prioritized drinking water policy has improved the extent of coverage of water networks. Thus it is hypothesized that an improvement in coverage of water supply network would have resulted in an increase in income by way of water charges.

Table 12.8: Details of connections in municipalities under the study (2002)

Particulars	*Municipalities*				
	Kheralu	*Siddhpur*	*Vadnagar*	*Unjha*	*Palanpur*
Total connections/total households (per cent)	72.0	96.5	62.1	70.7	100.0
Drinking connection/total households (per cent)	94.4	98.1	97.7	98.5	93.8
Population under piped network/ total population (per cent)	92.7	93.3	96.0	97.2	98.8

Source: Field survey (2003-04).

Major Income is Earned from Water Supply

Thus, on examining the data on income earned by Palanpur municipality, it was found that income from supplying water forms a major chunk of the total income of the municipality (Table 12.9). Between the years 1998-99 and 2003-04, it can be seen that an average of nearly 98 per cent of the total income earned by the municipality was earned from supplying water (Table 12.9). This shows the importance of income earned by supplying water in the total revenue of municipalities.

Cost Incurred

Table 12.10 shows the percentage of various costs incurred in supplying water. In the year 2003-04, cost incurred on salary was 70 percent of the total cost. Electricity charges, which accounted for nearly 35 per cent of the total costs incurred on water supply in the year 1997 reduced to just 3 per cent in 2002 (Table 12.10). This decline can be largely attributed to the availability of surface water from Dharoi reservoir. Reduced dependence on groundwater meant that less water was required to be pumped out and this resulted in substantial savings in electric charges (Table 12.10). However dependence on surface water rather than groundwater increased O and M cost of supplying water. This has shown an increase from 10 per cent in the year 1998-99 to 30 per cent in 2003-04 (Table 12.10).

Table 12.9: Income and cost incurred (at constant prices) in supplying water by Palanpur municipality

Particulars	*Years*					
	1998-99	*1999-00*	*2000-01*	*2001-02*	*2002-03*	*2003-04*
Actual income earned by municipality (Rs. per 10,000 litres)	446	400	245	196	494	446
Income earned if water charges were fully recovered (Rs. per 10,000 litres)	1899	1914	2144	1701	1395	1899
Percentage of actual income earned	95.1	94.5	94.4	94.3	98.2	97.5
by supplying water to total income earned by municipality (per cent)						
Percentage of actual income earned by municipality to income if water charges were fully recovered (per cent)	23.5	20.9	11.4	11.5	35.4	23.5
Percentage of actual income earned by water charges to cost of supplying water	13.0	7.9	3.4	3.7	33.4	23.8
Percentage of income earned from water charges if full recovery was collected to cost of supplying water	55.2	38.0	30.0	31.8	94.2	62.6
Cost incurred						
Cost of supplying water (Rs. per 10,000 litres)	3441	5041	6329	5069	1481	3441
Difference between costs of supplying water and income earned by municipality						
Deficit of income in situation when actual income earned by municipality from all sources were considered (Rs. per 10,000 litres)	2995	4640	6085	4873	987	2995
Deficit of income if full water charges was recovered (Rs. per 10,000 litres)	1542	3127	4185	3368	86	1542

Source: Field survey (2003-04).

Table 12.10: Proportion of costs to total costs incurred in supplying water by Palanpur municipality

Year	*Per cent Cost to Total Costs Incurred in Supplying Water*					*Total Water Cost (Rs. in millions)*
	Salary	*Electricity*	*Operation and Maintenance*	*Tanker*	*Treatment*	
1998-99	51	35	10	3	0	8.6
1999-00	30	45	25	0	0	13.7
2000-2001	27	40	21	0	0	15.3
2001-2002	27	3	34	0	37	18.9
2002-2003	70	0	27	0	4	6.4
2003-2004	70	0	30	0	0	9.6

Source: Field survey (2003-04).

Poor Recovery of Water Charges

In order to understand the financial viability of municipality supplying water to its people, this paper analysed the gap between costs incurred in water supply and revenue earned by supplying water. Table 12.9 shows that Palanpur municipality faces deficit in revenue account. A substantial loss was incurred in failing to collect water charges from all users. Table 12.9 indicates that recovery of water charges is poor.

Income and expenditure items were analysed for each year for the past six years from 1998-99 to 2003-04. It was evident from Table 12.9 that recovery of water charges declined for the first three years from 1998-99 to 2000-2001. During these years, municipality was dependent on groundwater for its water supply. Per capita water supply from municipality to its users throws light on the fact that groundwater failed to provide adequate water supply to meet the needs of the people. Thus, inadequate availability of drinking water has failed to recover water charges. The figures for the last three years showed the data after the commencement of water supply scheme (Table 12.9). With the improvement in coverage and extension of water supply network in the post 2001, water from the reservoir complemented groundwater source and resulted in improvement in water availability. Table 12.11 showed that with an improvement in water availability (after the year 2000), recovery and subsequently the income from water supply improved. This improvement in income can be attributed to assurance of adequate water supply. It can be safely concluded that inadequate water supply acts as a disincentive to getting the users to pay for the water consumed while adequate supply acts as an incentive for people to pay for water supply.

Table 12.11 reveals that prior RWSS, in the year 1998, the actual amount earned on water charges covers only 24 per cent of the cost of supply of water. In post RWSS phase (after 2001), water charges collected by this municipality increased drastically in the subsequent years (Table 12.11). The amount of actual income earned in the year 2002 was the highest. Even in this year, actual highest amount recovered from users is only 35 percent of the amount that could have been earned had the full water charges been recovered (Table 12.9).

However, all the attempts to manage the supply have failed to fully recover water charges (Table 12.9). Table 12.9 showed that even if water charges were to be recovered fully; they would be able to cover a maximum of 63 percent of the total cost of supplying water. Similar argument can be made by comparing cost and income earned by supplying 10,000 litres of water to their users. Again in year 2003, it was clear that even if full water charges were to be recovered, municipality would incur a loss of Rs. 1542 per every 10,000 litres of water supplied. Table 12.9 indicates that real income earned from supplying water also follows a declining trend as there were no efforts made to either recover dues or raise water charges.

Constant Water Charges over the Years

In per capita terms, income from water marginally increased from Rs. 2.17 in the year 1998-99 to Rs. 2.23 in the year 2002-03 after showing a decline in the three consecutive years (Table 12.11). Despite increasing the coverage and improving water supply, if the income falls then it could be attributed to low water charges.

Table 12.11: Income and costs incurred per capita (at constant prices) in supplying water by Palanpur municipality

Details	*Years*					
	1998-99	*1999-00*	*2000-01*	*2001-02*	*2002-03*	*2003-04*
Per capita income earned from water supply	2.23	1.92	0.90	0.91	2.17	1.14
Per capita income earned by municipality	9.31	8.92	8.64	7.83	8.51	5.30
Per capita cost of supplying water	16.81	23.30	26.72	24.41	6.70	8.51
Per capita loss incurred if income from water supply is considered	-14.61	-21.50	-25.72	-23.52	-4.52	-7.33
Per capita loss incurred if total income earned from municipality is considered	-7.51	-14.51	-18.00	-16.71	1.93	-3.22

Source: Field survey (2003-04).

Mounting Losses and an Explanation

Table 12.11 showed that the expenditure incurred in supplying water in per capita terms was Rs. 17 in constant terms in 1998. This increased to Rs. 27 in 2001-02, During this year, losses incurred by municipality in supplying water increased from Rs. 15 to Rs. 26 in constant terms (Table 12.11). After the year 2001, when the work on water networks was completed, Palanpur town became dependent on water from the reservoir. The gap between income and expenditure would have been more if one were to include capital works. This gap could be explained as follows:

1. *Under-Pricing*: One of the major reasons of this phenomenon is the under-pricing of water. A nuance of the phenomenon is that the residents have been enjoying a huge but hidden subsidy from their water supply.
2. *Low recovery of water charges*: Another reason is low recovery of water charges. If the municipality had been able to recover full water charges from its users, the per capita losses in the year 2003 would have reduced to Rs. 7 in constant terms (Table 12.11).

Very little left for Operation and Maintenance and its Consequences

The losses in constant terms indicate that little amount is left to carry out operation and maintenance works. This will have an adverse effect on the quality of service provided by the local body. Similar observations were made by the officials. They stated that municipality is able to recover only 15-20 per cent of the total water charges from its users. Though operation and maintenance costs of water supply increased from 10 per cent in 1997-98 to 30 per cent of total cost in the year 2003-04, low recovery leaves little funds to carry out O and M works (Table 12.8). Fortunately, there is a modest demand for O and M expenditure presently as the water works are newly constructed. Right now more important is to carry out O and M work on drainage pipelines. However, if operation and maintenance is not carried out, there could be grim consequences. Some of the obvious ones are:

1. *Water Contamination due to Leakage in Drainage Pipelines*: Officials also raised concern over the old drainage pipelines of Palanpur municipality, which have become dysfunctional. Water contamination due to leakages is obvious at places where drainage lines are laid in proximity to water supply pipelines. Officials believe that maintenance work on drainage pipelines is not carried out because of insufficient funds.
2. *Dependence on Other Institutions*: Dearth of funds to carry out O and M work increases the dependence of municipalities on external sources to carry out these works. The external sources include loans and grants from the state government and international financial institutions.

The only way out of this impasse is an improvement in water recovery charges. The above discussion shows that even in situations where attempts were made to improve adequacy and reliability of water supply, local bodies were unable to generate sufficient incomes. One of the reasons for involving users in water distribution and management by forming a municipality was to bring about an improvement in distribution of water and collection of water charges. Despite efforts in this direction, there is no visible impact on the financial conditions.

Constraints and Ways Forward

Summing up, results of this analysis show the relative performance of municipalities, as also the intra-municipal dependence and conflicts. Adequacy of water has no doubt improved but it is mandatory to view such an improvement in the perspective of under-utilised capacities of newly constructed water networks. This raises concerns about the sustainability of this institution.

There appears to be an increasing emphasis on the part of the department to supply water on the basis of BWR. In order to achieve the norms, supply-augmentation schemes are adopted. In this situation, water providers have little desire of transition to reforms based approach leading to efficient water use for meeting future needs. Moreover, municipalities suffer from problems of low water charges and poor cost recovery. This has adverse effect on maintenance of existing infrastructure, thus wasting water. The above analysis shows that standards achieve nothing unless they can be implemented and enforced. When establishing national standards, consideration must be given to the practical measures that will need to be taken with respect to finding new sources of water supply, providing for adequate surveillance and enforcement and managing demand by way of improving recovery and raising water charges to cover operation and maintenance expenditure in supplying water.

The proper analysis of pricing requires well-maintained records on quantity of water supplied for residential use and commercial use and the income earned from various users within the municipality. Such records are currently non-existent. In absence of much information, municipality is in no position to devise strategies to distribute water between uses or revise water rates.

We feel that performance of water use by municipalities can be improved by involving all stakeholders like department officials, officials of municipalities, beneficiaries and so on. There is a need to build capacities of municipal officials, train stakeholders and youths and motivate them through involvement in development programmes. This calls for the need for collective action in distributing and managing water supply. Some of the positive initiatives and success stories about community participation and cost recovery, which are observed to a limited extent in irrigation sector in the state and in drinking water sector in other states, do raise the hope of creating an economically viable and financially self dependent water supply sector in the state. The challenge is how to replicate the sector or region specific success stories. Efficient and effective uses of all water resources in drinking water use require the formulation and implementation of programmes for cost recovery, institutional reforms, laws and legislations and user participation. This requires a major shift from the paradigm stating government control over water resource planning and management to an innovative paradigm of users participation in operation and maintenance of water networks. Such a step is necessary to bridge not only the financial gap but also the performance gap afflicting water supply in the municipality.

This means that though water use in drinking water occupies only a small proportion of demand for water from the reservoir in comparison to farm use, it has the potential for competing with agricultural water use for the following reasons.

- ☆ First is that during years of scarcity this miniscule demand actually claims a major share of the total water available in the reservoir.
- ☆ Second is increasing urbanisation and depleting groundwater table increase the demand for drinking water significantly thereby increasing pressure on available water resource.

References

Biswas, Asit K. (1998). "Water Development and Environment." In Biswas (ed.) *Water Resources: Environmental Planning, Management and Development*, McGraw-Hill, New York.

Pathak, Jharna (2006). 'Irrigation Management Transfer: Problems and Policies' Unpublished Thesis submitted to the Gujarat University, Ahmedabad, Gujarat.

Rosegrant, M. W. and H. Binswanger (1994). 'Markets in Tradable Water Rights: Potential for Efficiency Gains in Developing Country Water Resource Allocation'. *World Development*, **22(11)**: 1613-1625.

Seckler, D., U. Amarasinghe, D. Molden, R. de Silva, and R. Barker (1998). 'World Water Demand and Supply, 1990 – 2025: Scenarios and Issues'. *IIMI, Research Report No. 19*, Colombo, Sri Lanka.

Young, R.A. (1996). "Measuring Economic Benefits for Water Investment and Policies," World Bank Technical Paper, No: 338. Washington D.C.

Management Strategies

13

Agroforestry vis-à-vis Environment and Disaster Reduction

A.S. Gill

INTRODUCTION

India is predominantly an agricultural country as more than 70 per cent of its population lives in villages and depends mainly on agricultural and animal husbandry. Out of the total geographical area of 329 M ha, only 142 M ha is meeting out the demand of one billion populations. Besides meeting the food grain needs, the agricultural needs of 676 tones of dry forage for cattle and 46 million m3 of fuelwood by 2020 are to be met. About 57 per cent of the total geographical area of the country (about 187 M ha) has been affected by various land degradation mainly through human made problems and it's over exploitation.

The Forest Survey of India (1995) estimated that the area under the forests changed from 64.01 to 63.96 M ha between 1993 and 1995. The maximum loss of area was in northeastern states where the area under open forests decreased and under closed forests it increased. The recorded forest area is 76.5 million ha or 23.3 per cent of the total geographical area of the country. At present the actual forest cover is estimated to be 63.34 million ha *i.e.* 19.27 per cent of country's area.

It is estimated that the population of India will increase to about 1.44 billion by 2030 AD necessitating commensurate increase in the production of food grain. The total livestock population has increased from 445 M in 1987 to 470 M in 1992. This high increase in human and livestock population in the country has caused a heavy pressure on the land, water and air, resulting in environmental problems in addition due to urbanization/industrialization process.

Agroforestry although a new term, coined in 1977, it denotes an age-old practice of having trees mixed in the agricultural landscape. Agroforestry research in India is of recent origin though the ancient Indians practiced agroforestry. The practice of allowing of trees of Khejri (*Prosopis cineraria*) in arid region and *Acacia* sp. in the semi-arid regions is age-old practices of agroforestry in India. Similarly, the cultivation of large cardamom (*Ammomum subulatum*) in combinations with *Alnus nepalemsis* in Central Himalayas, various tree crops combination in Jhoom area of NEH region is being practices since decades.

Environment

Environment comprises of land (soil), water and air. If the environment is polluted then there will not be a good food to eat good air to breath and good water to drink. This may be in the case of human, animal and the plants. Though the law of land regarding protection of the environment is very stringent, but how these laws are implemented is more important.

Land

Out of the total geographical area of 329 M ha, about 187 M ha is facing one form or the other degradation problems (water erosion, wind erosion, chemical deterioration and physical deterioration has resulted soil degradation in an area to the extent of 48.9, 13.5, 13.8 and 11.6 M ha, respectively).

Water

Large area in the country is still rainfed and even if all resources are pooled even than it will remain more than the 50 per cent under rainfed. Normally, with the change in agro-climatic conditions the weather is unpredictable. Large areas are prone to drought each year and agriculture is thus becoming un-economical day by day. Mainly this is being focused on the declined trend in the forest cover. Also in the irrigated area, its water quality has gone down considerably mainly due to heavy pollution of the rivers primarily due to city waste and industrialization. The canal's supplying irrigation water is full of weeds and the hyacinth problem. Due to heavy pressure on irrigation water due to repeated droughts and greater demand, the groundwater level is receding at a faster rate. Even in a state like Punjab this is causing great concern to the planners. For billions of people, access to safe drinking water will remain a mirage as domestic sewage and industrial wastes contaminate river and wells.

Air

It may constitute just a fraction of the atmospheric gases: 0.03 per cent of the total. But without carbon dioxide our planet would be as cold as the moon. With the in increase of CO_2 in the atmosphere, we are recording high temperatures. CO_2 concentration is increasing by 2 per cent year to year, which will have a direct influence on patterns of atmospheric circulation, polar-ice shields at sea levels for precipitation and its distribution, agricultural forest productivity and CO_2 flux between the biosphere and the atmosphere of our planet. It is stated primarily due to coal burning and vehicles running on the roads. Reduce in the forest cover is also to be blamed.

Scientists predicts the increase one to two degrees of the earth temperature in next two decades (in some cases it may go up by six degrees), since ocean act as reservoir of heat, it could result in rising sea levels.

Agroforestry Systems

The definition, which has gained the widest international acceptance, is "agroforestry is a collective name for all land-use systems and practices where woody perennials are deliberately grown on the same land management unit as agricultural crops and/or animals, either in spatial mixture or in temporal sequence". There must be significant ecological and economical interactions between the woody and non-woody components.

An ideal agroforestry system should possess the three attributes (Nair 1984):

1. Productivity
2. Sustainability and
3. Adaptabiltry

Basically in all agroforestry systems there are two common roles:

Productive Roles

In a given agroforestry system there are two or more production like food, fodder, fuel, fibre, fertilizer and fruit, etc. Commonly, agroforestry system is popularly known for "five F".

Service Roles

Agroforestry systems were mainly borne for its service roles.The major ones being soil and moisture conservation, fertility improvement, water and wind erosion control, extending shade and shelter, etc.

Numerous agroforestry systems have been developed in different areas. Gill and Bisaria (1997) suggested following common agroforestry systems in India on the basis of nature of components.

1. Agrisilviculture (Trees + crops); 2. Boundary plantation (Trees on boundary + crops); 3. Block plantation (Trees + crops); 4. Energy plantation (Trees + crops during initial years); 5. Alley cropping (Shrubs + crops); 6. Agrihorticulture (Fruit trees + crops); 7. Agrihortisilviculture (Trees + fruit trees + crops); 8. Agrisilvipasture (Trees + crops + pasture/animals); 9. Hortipasture (Fruit trees + pasture/animals); 10. Silvipasture (Trees + pasture/animals); 11. Forage forestry (Forage trees + pasture);12. Shelter belts (Trees +/- crops); 13. Wind break (Trees +/- crops);14. Live fence (Shrubs/ under trees on boundary); 15. Homestead (Multiple combination of trees, fruit trees, vegetable, etc.); 16. Entomoforestry (Trees + sericulture); 17. Aquaforestry (Trees + fishes)

In the various agroforestry systems the dominant production and service roles are played by the multipurpose trees species. The major multipurpose tree species for the various regions of the country are reported in Table 13.1.

Table 13.1: Multipurpose tree species for different agro-climatic regions of India

Agroclimatic Regions	*Prioity*				
	1	*2*	*3*	*4*	*5*
Western Himalayan	*Grewia optiva*	*Populus Ciliata*	*Toona ciliata*	*Casuarina australis*	*Acacia catechu*
Eastern Himalayan	*Michelia champaka*	*Alnus nepalensis*	*Gmelina arborea*	*Morus laevigata*	*Pinus kesiya*
Lower Gangetic Plains	*Eucalyptus hybrid*	*Acacia auriculiformis*	*Gmelina arborea*	*Acacia nilotica*	*Azardirachta indica*
Middle Gangetic plains	*Populus deltoides*	*Anthocphalus cadamba*	*Eucalyptus hybrid*	*Dalbergia sissoo*	*Acacia nilotica*
UpperGangetic Plains	*Populus deltoides*	*Eucalyptus hybrid*	*Dalbergia sissoo*	*Anthocephalus cadamba*	*Leucaena leucocephala*
Trans-Gangetic Plains	*Populus deltoides*	*Eucalyptus hybrid*	*Dalbergia sissoo*	*Melia azadirach*	*Acacia nilotica*
Eastern Plateau and Hills	*Gmelina arborea*	*Tectona grandis*	*Eucalyptus hybrid*	*Casuarina equisetifolia*	*Leucaena leucocephala*
Central Plateau and Hills	*Azardirachta indica*	*Eucalyptus hybrid*	*Tectona grandis*	*Acacia nilotica*	*Leucaena leucocephala*
Western Plateau and Hills	*Azardirachta indica*	*Acacia nilotica*	*Eucalyptus hybrid*	*Leucaena leucocephala*	*Tectona grandis*
Southern Plateau and Hills	*Ailanthus excelsa*	*Eucalyptus camaldulensis*	*Tamarindus indica*	*Cieba pentandra*	*Casuarina equisetfolia*
East coast Plains and Hills	*Casuarina equisetifolia*	*Gmelia arborea*	*Acacia mangium*	*Tectona grandis*	*Dalbergia sissoo*
West Coast Plains and Ghats	*Casuarina equisetifolia*	*Eucalyptus hybrid*	*Acacia mangium*	*Terminalia tomentosa*	*Artocarpus heterophyllus*
Gujarat Plains and Hills	*Prosopis cineraria*	*Eucalyptus hybrid*	*Ailanthus exelsa*	*Dalbergia sissoo*	*Leucaena leucocephala*
Western dry	*Prosopis cineraria*	*Acacia nilotica*	*Azardirachta indica*	*Ailanthus excelsa*	*Dalbergia sissoo*
The islands	*Casuarina equisetifolia*	*Gmelina arborea*	*Gliricidia sepium*	*Smania saman*	*Terminalia catapa*

Source: ICAR Vision 2020.

Multipurpose Tree Species

Trees are a vital component of the survival economy of the rural poor. Their life is dependent on trees. They collect wood to cook their meals and tree foliage for their animal. Tree alongwith soil play a crucial role in the global cycling of carbon, the importance of which has beam magnified by the emergence of carbon dioxide induced climatic change as arguably the most threatening environmental problem of modern times (Postel and Heise 1990). Trees have a large and indispensable role to play in improving present and future food security. Trees play a critical role in ensuring sustained agricultural production, including animal husbandry and in some instances fisheries. Thus, for physical, economic and social reason development strategies need to be aimed at total production and environmental protection through promoting horizontal integration of production. Under these circumstances there is hardly any other alternative but to promote integrated agroforestry system with the various multipurpose tree species.

Nitrogen Fixing Tree Species

The trees especially nitrogen fixing trees species (NFT's) play a vital role in environmental protection. The NFT's grown on marginal/sub-marginal/degraded lands particularly under agroforestry systems help in environmental stability by reducing land degradation, keeping atmosphere free of pollutants, moderating temperature, etc.

About 650 trees and large shrub species belonging to 9 plant families are known to fix nitrogen (Brewbaker *et.al.*, 1990). The most of them (80 per cent) are legumes (Leguminosae) with 320 Mimosoids, 170 Papilionoids and 25 Caesalpiniods. The non-legume families are Betulaceae, Casuarinaceae, Cariariaceae, Elaeagnaceae, Myricaceae, Rhamnaceae, Rosaceae and Ulmaceae. The most important NFT's genera are *Acacia, Albizia, Alnus, Calliandra, Casuarina, Dalbergia, Gliricidia, Leucaena, Prosopis, Robinia, Sesbania,* etc. The trees fix the atmospheric nitrogen as a result of symbiotic association with Rhisobial bacteria in case of Leguminosae and in some member of Ulmaceae, while in other NFT's actinomycete genus Frankia is involved in N-fixation. The Rhizobia or Frankia provide amino acids for combined organism. Three genera of Rhizobia namely Rhizobium, Bradyrhizobium and Azorhizobium are involved in N-fixation in NFT's. The microbial symbiont infects the roots of NFT's and stimulates the tree to form nodule. With in the nodule the symbiont multiplies and fixes nitrogen. The tree contributes to the process of symbiosis by providing the symbiont with suitable environment for N-fixation and energy in the form of carbohydrate from photosynthesis.

Factors Effecting N-fixation by NFT's

Climatic variation and soil factors limit the N-fixation by NFT's. Some of them are discussed below.

Temperature

Generally, the Rhizobia are more sensitive to high temperature. Exposing inoculated tree seeds even a short period of time at high temperature can result in death of bacteria and reduction in nodulation. However, different strains of Rhizobium respond differently to temperature variations.

Oxygen

The abundant supply of oxygen is necessary for formation and functioning of nodules.

Water Stress

Too low or too high soil water content limits the process of N-fixation.

Soil Acidity

Soil acidity generally reduces N-fixation by inhibiting the development of Rhizobium, increasing the numbers of ineffective Rhizobium or disrupting the infection processes (Sprent and Sprent, 1990; Mengel and Shubert, 1988).

Soil Salinity

The high salt content of saline soil can inhibit N-fixation by osmotic withdrawal of moisture for nodules, decrease in bacterial colonization of roots and root hair curling

Environment Protection by NFT's

Land Degradation

The recognized forms of soil degradation are physical, chemical and biological degradation. The physical degradation including soil erosion (by water and wind), desertification, compaction and hard setting. Chemical degradation includes acidification and sodification, biological degradation covers decline in organic matter and reduction in soil biological activity. The potential of trees has been recognized and given new focus to reclaim and restore degraded lands.

Environmental Pollutants: Greenhouse Gases

The atmospheric concentration of CO_2 and other gases have been steadily increasing over the past century or more (Batch, 1986) causing changes in global climate. The modern agriculture involving various land-use practices *e.g.* burning, fertilization, frequent disturbance and tillage etc cause emission of green house gases. The intensive cultivation, therefore, can result in globally significant emission of green house gases (Dixon *et al.*, 1994). The conversion of forest or grassland to intensive agriculture generally results in decrease in soil organic matter and increase in atmospheric CO_2 due to increase rates of organic matter decomposition and microbial respiration. Keeping in view the adverse effect of intensive agriculture in environmental degradation alternative land-use systems should be followed that can increase carbon content in soil and vegetation by increasing the amounts of sequestered and stored carbon dioxide, and reduce the emission of green house gases. The growing of trees especially NFT's in various agroforestry systems can act as sinks of green house gases and sequester carbon in terrestrial system. This low-input management system, which minimize soil and plant disturbances, maintains perennial vegetation cover and emphasizes nutrient cycling, that contribute to storage pools of carbon and nitrogen that are stable over decades or centuries. The agroforestry systems that are evaluated in 50 nations revealed various practical ways of reducing green house gas emission and conserving or sequestering carbon

Important Agroforestry Systems in Aid of Environmental Management

Agroforestry systems are for sustenance of environment (Gill and Bisaria 1997) and the important areas where agroforestry systems maintain and improve the environment are 1.Air pollution; 2.Noise pollution; 3.Precipatation; 4.Temperature and 5. Wind velocity.

The important agroforestry systems known for environment management are: -

Alley Cropping

Alley cropping is a farming method in which food crops are planted in alleys between rows of fast growing leguminous trees or shrubs (Gill 1998). The trees or shrubs require some pruning during cropping to avoid excessive shading and competition with the food crops in the alleys. This system was developed to replace shifting cultivation and bush fallow systems and, to promote soil and water conservation on slopping land or hilly areas (Gill 1995). Through agroforestry systems, steep slopes can be managed effectively in the control of soil erosion (Gill and Lal 1995).

Wind Breaks/Shelter Belts

These are belts/blocks consisting of several rows of trees established at right angles to the prevailing wind. Windbreaks and live fences are options available in agroforestry for dry areas. In desert large planting trees have stabilized areas as wind breaks. Wind breaks/shelter belts reduce soil erosion and there is sufficient information available where grain yields were not reduced to the desired level. China has shown the real importance of Shelterbelt. The "Three North" Shelterbelt system was evolved to protect the existing forests, grasslands and crops, to establish the sand fixation forests water and soil conservation forests.

SALT

It is known as Sloping Agricultural Land Technology (SALT) as developed by non-governmental agencies in the Philippines. The focus is on sloping land, where many small-scale farmers face serious problems of soil erosion. Its main objectives are to control soil erosion; help restore soil structure and fertility and to produce food efficiently.

Conservation Farming

It is a farming where land holdings are small. It has special significance to India where land holdings are small especially in the hilly states. Farmers would gain considerably if they adopted conservation farming for their landholdings and it would then yield more without added expense. Conservation farming takes the self-regenerating forest as its model. Fast growing nitrogen fixing tree species are chosen which fix nitrogen in the soil and crops are sown in between (similar to alley cropping).

Rainfed Agroforestry

Under rainfed conditions the climatic condition are harsh and soil undulating and low in fertility status. The over all environments are not conducive to farming. Agroforestry intervention makes agriculture quite safe. Rainfed agroforestry results in improving and stabilizing the income of the farmers (Roy and Gill 1991; Gill et al 2004). Boundary plantations of nitrogen fixing tree species have borne fruitful results (Gill 2000).

References

Batch, W., 1986. The gases and their influence on climate. *Natural Resources and Development*, **24**: 90-124.

Brew baker, J.L., Willers, K.B. and Macklim, N., 1990. Nitrogen-fixing trees: Validation and prioritization. Proceedings IUFRO Congress, Aug., Montreal.

Dixon, R.K, Winjum, J.K., Andrasko, K.J., Lee, J.J. and Schroeder, P.E., 1994. Integrated systems: assessment of promising agroforestry and alternative land-use practices to enhance carbon conservation and sequestration. *Climatic Change*, **30**: 1-23.

Gill, A.S. and Bisaria, A.K., 1997. Agroforestry for sustenance of agriculture and global environment. In. *Agro's Annual Review of Crop Ecology* (Eds. A.K.Dahama). Agro Botanical Publisher, Bikaner, pp. 55-78.

Gill, A.S. and Lal, B., 1995. Erosion control in steeplands through agroforestry. In. Proceedings International Symposium (EDT's. R.B.Singh), Delhi University, Delhi, pp. 451 -475.

Gill, A.S., 1998. Alley cropping for food fodder production. *Indian Farming*, **48(2)**: 27-29.

Gill, A.S., 2000. Food grains production with planting trees on field boundaries. *Indian Farming*, **50(5)**: 36-39.

Gill,A.S., Gupta, S.K. and Handa, D.P., 2004. Rainfed agroforestry system in semi-arid areas. *SAIC Newsletter*, **14(1)**: 6.

Mengel, K. and Shubert, E., 1988. Effect of soil pH on nitrogenase activity, crop yield and *Rhizobium* multiplication. In: *Physiological Limitations and the Genetic Improvement of Symbiotic Nitrogen Fixation*, (Eds.) F.S. O'Gara, S. Manian and J.J. Drevon. Kluwer Academic Publishers, Dordrecht, pp. 137-46.

Nair, P.K.R., 1984. Soil productivity aspects of agroforestry, ICRAF, Nairobi, pp. 85.

Postel, Sandra and Heisae, Lori, 1990. Reforesting the earth. *Span* June, pp. 11-15.

Roy, Debroy and Gill, A.S., 1991. Agroforestry systems as a means for minimizing risk in rainfed agriculture. In. *Tech. for Minimizing Risk in Rainfed Agric.* (Eds.) S.P. Singh and C. Prasad. Indian Society of Extension Education, New Delhi, p. 156-159.

Sprent, J.I. and Sprent, P., 1990. *Nitrogen Fixation Organisms*. Chapman and Hall, London.

14

Ecosystem Approach to Resurrection of Gulf Coast

Ashwani Vasishth

Natural Processes, Disastrous Consequences

Disasters are conventionally thought to be catastrophic events. And they are that, in many ways. Certainly, they command our attention. However, I argue that these events we call disasters are situated within a broader context of complex bio-geo-chemical processes and functions, rather than merely as singular space-time events. Besides, disasters are best seen as integral to human systems, rather than external to me. Then, truly effective planning interventions intending to anticipate and mitigate disaster events need to be grounded in an adaptive ecosystem approach based on a scale-hierarchic process-function oriented ecology.

The root premise here is that the manifest events and entities which comprise physically tangible natural reality are best thought of as the end-points of complex bio-geo-chemical processes and functions. In other words, reality is actually constituted by processes and functions which are often not directly perceivable by human senses, rather than by the aggregation of tangible events, entities and occurrences. What we are able to directly sense are the morphologically manifest events and entities, which are incomplete proxies of the actual processes and functions that shape reality. Then, what we see and taste and touch and hear and smell, and thus take to be real, are in fact abstractions from an occurring actuality. We need to see beyond the events and entities directly perceivable by us to the processes and functions that underwrite them. I argue that our planning interventions must be shaped more by these often intangible processes and functions than by the morphological events and entities of which we are more directly and empirically aware.

Natural phenomena become disastrous to us by virtue of their impact on our world. These disasters are real enough, and matter most to us because of their direct impacts on our material world. But the wildfires, typhoons, tsunamis, earthquakes, floods, and cyclonic storms should be seen as necessary natural processes to which we must adapt, rather than horrid events that must be circumvented or somehow thwarted. At the human level of organization, they are clearly disastrous. But, taken at a broader level of organization, they are merely occurrant.

We discover, for instance, that fire suppression is not a solution to disastrous wildfires, but rather itself the source of new problems that factually make wildfires even more disastrous. When we suppress wildfires, we change understorey vegetative growth patterns, and tree densities that add fuel, so to speak, to any subsequent fires that might erupt. Differently, disasters are scale events. In the sense that, at the scale of a few tens of square miles, a wildfire is categorically disasterous. But the same wildfire, taken at the scale of a few thousand square miles, is merely a necessary disturbance.

Earthquakes and tsunamis are incidental outcomes of the tectonic forces that shape the terrain upon which we live. The Earth, and all that it embodies, could not have taken shape absent these necessary forces. Hurricanes and typhoons are merely manifestations of meteorological and atmospheric processes to which we must learn to adapt and live around. In short, we must learn to live with disasters, rather than strive to make them unhappen.

It is my contention that effective interventions which might help us to live with disasters, and which have the potential to reduce the physically tangible impacts such events have on our experienced world, must be shaped most by the processes and functions that drive these disastrous events, rather than merely upon some representation of the events themselves. Our response to disaster must be adaptive to, and mitigative of, the salient processes and functions. Our responses must show our ability to cohabit with reality rather than somehow to conquer it. The cyclonic storms that ravage the Eastern seaboard of the Indian sub-continent at least twice a year, the wildfires that rage seasonally through the landscapes of Southern California, the earthquakes that send shock and tidal waves surging across the land, these are all manifestations of processes and functions that shape our world, rather than merely catastrophic and terrible events that we would seek to wish away. In short, most events that we call disastrous are natural events to which nature itself provides us with mitigative adaptations that might help us to "ride out the storm," so to speak.

More importantly, we find that we often make matters worse for ourselves, by disregarding the very processes and functions that might help us to live less disastrously with these natural phenomena. In the case of wildfires in North America, for instance, we have so effectively suppressed fires upon the landscape that the forests have grown to a tree density four times the norm that would have existed if fires had been allowed to run their course, thus making fires that do happen much more disastrous.

The destruction of mangrove swamps and coastal wetlands pursuant to our shortsighted desire to capitalize on the scenic beauty of shoreline living makes the actual impact of storm events that much more disastrous. And the reengineering of rivers to make them more amenable to human economic need as navigable channels may end up diverting sedimentation processes that would normally have formed coastal barrier islands that might well have protected us from the worst ravages of this hurricane or that.

Then, intelligence requires that we understand and abide by the processes and functions that shape the events we think to be disasters, and learn to live adaptively with these events by understanding the processes and functions that shape them. In order to do this well, we must learn to think ecologically about context and about consequence. We must teach ourselves to fully appreciate contextual complexity, even as we reduce our world to some more manageably simplified model of reality. And we must learn to trace out the consequences that cut across levels of organization in cascading, nested chains of cause and effect. Instead we live most often oblivious to these processes and functions, shortsightedly casting the world into forms that better suit our short-term economic needs, and not thinking through the ecosystem consequences of choices we make.

Nested Levels, Scale Dependence and a Process-Function View as the Basis for an Ecological Approach to Planning

The devastation wrought by Hurricanes Katrina and Rita on the Gulf Coast cities, particularly New Orleans, is also a useful example in demonstrating the value of using an ecosystem approach from process-function ecology, partly by underscoring the value of ecological processes that we have conventionally excluded from our planning descriptions (at great subsequent cost to local and regional communities), even as we have treated planning itself as a discretely place-based activity, with singularizable boundaries.

Once we recognize complexity as a special case in planning requiring special conceptual and descriptive tools, as illustrated by Rittel and Webber [1973] and by Holling and Goldberg [1971], process-function ecosystem ecology readily presents itself as a protocol for dealing with the construction and management of instrumental depictions of this complexity. The key moves in such an ecosystem approach involve: an expansion of the well-entrenched but incomplete typological and population-community world-views into a broader *process-function view*; an adoption of the *nested systems* model of complexity, with its associated move to conceiving this complexity as being arranged into meaningfully named *levels-of-organization*; a recognition that complex systems reality is not readily singularized in any useful way, but must, instead, be described from strategically *multiple perspectives* using *multiple criteria*, with some explication of *purpose*; and an acknowledgement of the scale-dependence properties of complex system realities, requiring the considered application of *multiple spatial, temporal and organizational scales* in constructing system descriptions [Vasishth, 2008].

Evoking an Ecosystem Approach to Disaster Planning: Nested Systems, Multiple Functional Boundaries and the Ecological Recontextualization of New Orleans

A key principle from within an ecosystem approach to planning has to do with the imperative to use multiple temporal and spatial scales in conceptualizing context and tracing consequence. In planning, and in the case of multiple temporal scales, this principle is stymied by the practice of fiscal discounting, which is deeply ingrained into decision making protocols. This practice ensures that the future is always worth less–often much less–than the present. Partly as a consequence of this practice of fiscally discounting the future, and partly because many elected officials are strongly influenced by the time scales of their own election cycles, legislation-based and regulatory decision systems are afflicted by a form of attention deficit hyperactivity disorder that hugely handicaps functionally robust decision making and implementation. Such decisions are never able to take a sufficiently robust long-term view of the situation at hand, and, simultaneously, feel impelled to push hard for very brief periods of time, until attention is once more distracted by the movement of political event horizons.

This is especially true when such decision making needs to consider processes that play out at time scales dramatically different from those in the political process, or is driven by the equally short attention-span of popular news media's propensity to highlight crises in some fleeting way. The case of disaster planning, as underscored by the recent Hurricanes Katrina and Rita which struck the Lower Mississippi River Delta in August 2005, is an instance where an ecosystem approach would prove particularly beneficial.

The Imperative for an Ecosystem Approach to Disaster Planning

An ecosystem approach to integrative regional planning demands that good decisions in complex spaces cut across existing political and administrative jurisdictional boundaries, abiding instead by geographical, functional and ecological boundaries, and then playing out along more than two or three functionally relevant time lines.

Take, for instance, the case for protecting New Orleans from future storm-based flood events. The US Army Corps of Engineers (USACE) has noted that it is quite possible, technically, to protect the city from such storm surge events, using some combination of ecological restoration techniques that rehabilitate the wetlands and barrier islands which once naturally buffered the Lower Mississippi Delta, and a strategically constructed system of levees. For such protection to be put in place, however, the resources and the political will must be made reliably and predictably available over a time-scale of decades, rather than weeks or months [Anonymous, 2005].

Due to this attention-deficit hyperactivity disorder that afflicts public planning and political policy, we can be virtually certain that the complex set of actions necessary to buffer New Orleans from future storm events will probably not happen without conscious and sustained effort on our part. What is needed is some reliable way to change the practices of decision-making to ensure that appropriately multiple spatial, temporal and organizational scales are used in generating operational descriptions of ecological context and of consequence. A systems approach, based on scale hierarchic ecosystem ecology, offers a useful set of tools to help shape an alternative approach to planning, one that is able to take account of the multiple boundaries that shape complex realities, and the occurrant time-frames that underwrite ecological processes and functions.

In disaster planning, a central distinction is made between actions required in the relatively immediate aftermath of any particular disaster, such as providing rescue and recovery efforts, restoring transportation routes and communication channels, reintegrating and reprovisioning supply chains, providing survival relief, and providing trauma support, and the longer term restorative and preventative actions that need to be made to complete the regional recovery and reconstruction process and to integrate valuable lessons learned into institutional, technical and cultural practices. All this is vital to effectively mitigate against future similar events, and we detract from any one of these at significant costs to sustainability planning.

Context and Consequence, on the Lower Mississippi River Delta

Rivers are key organizing elements in watershed ecologies, dominating processes and functions and their arrangement across space and time, in a sense comprising the spinal column about which the body of its lands take shape, grow and change. The Mississippi River is no exception. New Orleans, and the areas that most immediately support its existence, are dominated by the reaches of the Lower Mississippi Delta. Wetlands and barrier islands are two key ecological elements that shape and structure that land, and the continued health and wellbeing of both are imperative to the future of the place we have come to value, both for its ecologies and for our human (economic and cultural) purposes.

Throughout history, but with increasing intensity during the more recent period of the land's settlement by white European immigrants and its subsequent appropriation into the United States of America, the river has been useful to the diverse ecological and human communities that have sprung up around its ebbs and flows. With increasing intensities of population, industrialization, and globalization pressures, the stresses and constraints imposed upon the riverine system have also increased.

The U.S. draws enormous economic benefits from this riverine Delta system. Examples of revenue sources include, farming of rice and other crops, fisheries resources extracted from the Delta ecosystem, nature tourism and hunting activities, direct natural resource extraction such as oil wells[2] and mining, the tourism-related benefits accruing from the cultural practices that have evolved within the bayou ecosystem, as well as in the river's use as a channel for the transportation of freight, linking global marine shipping with domestic U.S. goods movement into the American

---

2 Coastal Louisiana is the single largest domestic source of oil and condensate within the U.S. In addition, approximately 30 per cent of the nation's entire imported oil and natural gas supply is transshipped through the Louisiana Delta [USACE, 2004: ii]

heartland. Any useful description of rich context of this system would take account of each of the diverse sets of boundaries that each of these quite distinct purposes would project upon the ecological complex that is the Lower Mississippi River Delta.

The gradual but increasing channelization of the Lower Mississippi River has sought to constrain the natural tendencies of rivers to shift course in evolutionary response to landscape and process-function changes over time. The tendency of the river to overflow its banks seasonally, flooding the surrounding lands, resulted in the increased embankment of the Mississippi River. In addition, projects have sought to modify the course of the river in ways that minimize the costs of navigating the river from its Gulf of Mexico port facilities to the inland reaches of the U.S. hinterland.

Together, these actions have dramatically disrupted processes and functions ecologically vital to the reformation and maintenance of the wetlands that form the overarching context for New Orleans. These projects have also caused significant deteriorations of the barrier islands so integral to the ecosystem at large, and that have long buffered the city from Gulf storms. The US Geological Survey [USGS, 1995] warns that:

> (a)s the barrier islands disintegrate, the vast system of sheltered wetlands along Louisiana's delta plains are exposed to the full force and effects of open marine processes such as wave action, salinity intrusion, storm surge, tidal currents, and sediment transport that combine to accelerate wetlands deterioration.

The river carries down silt from its inland reaches, to be deposited throughout the wetland marshes as well as out into the Delta region, allowing the buildup of soils in a manner that counteracts the unceasing erosion of coastal lands by ocean currents, and the subsidence tendency of soils in marshy coastal regions. This process of siltification is an imperative that we have disrupted at our own peril, as Hurricane Katrina so vividly demonstrated for us.

Since the 1930s, coastal Louisiana has lost some 900,000 acres of coastal wetlands. "The current rate of loss is about 16,000 acres per year." [Galloway, 2003: 2] About 70 per cent of this loss in coastal wetlands has been estimated to derive directly or indirectly from human activity, and from our continued appropriation and alteration of the natural estuarine landscape [Izzo, 2004].

Ecological processes being driven by this human appropriation, and in addition to the loss of buffering coastal lands, barrier islands and wetlands, include subsidence and saltwater intrusions into aquifers due to draw-downs of groundwater, and the loss of other ecological and biological functional value deriving from habitat loss. To take just one example, migratory birds, the USACE [2004: ii] estimates that:

> (a)pproximately 70 percent of all waterfowl that migrate through the U.S. use the Mississippi and Central flyways. With over 5 million birds wintering in Louisiana, the Louisiana coastal wetlands are a crucial habitat to these birds, as well as to neotropical migratory songbirds and other avian species who use them as crucial stopover habitat. Additionally, coastal Louisiana provides crucial nesting habitat for many species of water birds, such as the

> endangered brown pelican. These economic and habitat values, which are protected and supported by the coastal wetlands of Louisiana, are significant on a National level.

Consider that at the time when New Orleans was established as a settlement by the French, in 1718, some 100 miles of barrier islands lay between it and the Gulf of Mexico. Since then, the Mississippi Delta has been losing land at the rate of between 25 and 50 square miles per year (which translates into approximately 2 acres per hour) [Barras *et al.*, 2003], in large measure because our efforts to maximize the immediate and short run economic benefits of the Mississippi River to humans in their present time.

> Before the levees were built to channel and "control" the Mississippi and other nearby rivers, floodwaters would spread out and slow down as they flowed over the delta. When the flow slowed, the river would deposit its burden of silt, forming a new layer of earth. But the levees, which now constrict floods along a 1,200-mile corridor of the Mississippi, keep the floodwaters from spreading across the delta. Instead, the river-borne silt is lost off the edge of the continental shelf. (Penland, 2005)

The Delta region now lays exposed to the elements in ways that will jeopardize human well-being long into the future. In very real terms, past human generations had captured economic wealth for themselves, while deferring the costs onto then-future (and now-present) generations. Perhaps because we continue to discount the future by skewing the time scales of our descriptions for economic analysis, and because we continue to exclude purpose and perspective in assessing the context and the consequences of proposed action, we are not doing much better, as planners.

Other natural and human forces are at work here as well, which must additionally be given standing in an ecosystem approach to disaster planning which seeks to better integrate context and consequence. In the case of global climate change, direct observations indicate that ocean temperatures are rising. This assures us that the intensities of future hurricane events are bound to increase. Similarly, sea levels have been observed to rise, in part due to the melting of glaciers, in part due to increased coastal runoff, and in part perhaps due to local land subsidence. In any case, the ways in which we choose to buffer ourselves from future storm events must take all these aspects into account in our depictions of context and our analysis of consequence.

Elements of an Ecosystem Approach to Disaster Planning

One essential and central element in such a restructuring must be some form of an ecosystem-approach to integrative regional planning, in a way that pragmatically changes how we make descriptions of problems, and how we think to set about addressing them. In a sense, the move is away from a "problem-solving" mode that seeks to eliminate the problem, or "make it go away," and toward an on-going and adaptive "problem-management" mode that seeks to engage context and consequence. A second and necessary element in such a move to adaptive ecosystem management has to do with the way we think of descriptions, and the ways information is

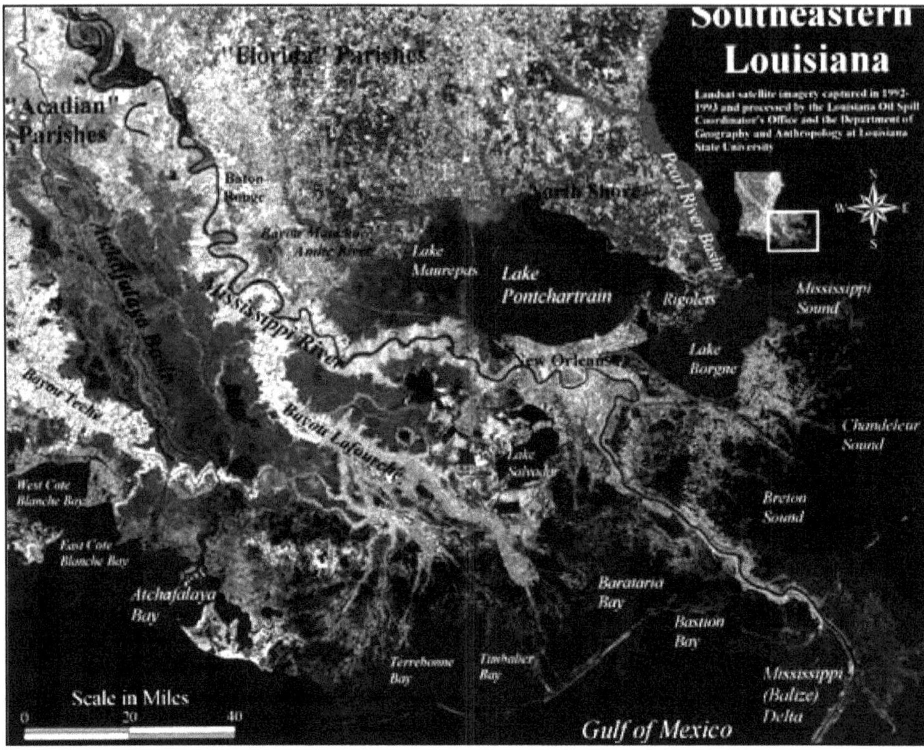

Figure 14.1: The southeastern Louisiana Barrier Islands as the ecological context for a New Orleans impacted by Hurricane Katrina.
[America's Wetland Resource Center, undated].

embedded within them.

Conceptualizing the world as being structured in the form of nested systems leads us to understand that those aspects of the world which demand description as complex systems must be conceptualized as having more than one or two functional boundaries. As a consequence, their depictions can only be constructed using multiple spatial, temporal and organizational scales. Evolutionary ecosystem ecology provides a sound basis for effecting such changes in our decision-support mechanisms, and in the development of long-lasting adaptive disaster management strategies.

Once again, Rittel and Webber's characterization of "tame" problems and "wicked" problems [1973], as a way of capturing key operational distinctions between simple, closed, mechanical systems, and complex, open, organic systems, is a useful place to begin our search for an adaptive ecological approach to disaster planning. When we are dealing with "wicked" problems, that is to say, problems emergent from the processes and functions that drive complex, organic and open systems, then descriptions must capture the multiple boundaries that structurally characterize complex systems if they are to be useful.

The diverse array of factors and forces that drive such an ecosystem approach to disaster planning, which involves setting the multiple boundaries of functional relevance, can only be made evident by the systematic and self-conscious use of appropriate spatial, temporal and organizational scales, coupled with an application of a levels of organization approach to system description. An ecosystem approach to disaster planning offers the necessary conceptual shift as well as the descriptive tools and techniques to counter the spatial and temporal biases inherent in conventional planning and policy–the singularization of place-based boundaries, and the fiscal discounting practices that skew temporal scales strongly toward the present.

Not only do we need to be able to think about place in more sophisticated ways, but the imperative for sustainability, when combined with the imperative of ecological complexity, requires that we use multiple boundaries and multiple functional scales in crafting our descriptions, if we are to capture complex reality in a way that more richly sets the context for our interventions, and shows us more of the potential ecological consequences of our actions.

References

Anonymous, 2005. "Protecting New Orleans from Future Floods: Billions of Dollars Over Decades," *CNN*, 17 September, 2005. Retrieved September 17, 2005, from <http: //www.cnn.com/2005/TECH/09/17/katrina.levees.ap/index.html>

Barras, J. *et al.*, 2004. *Historical and Projected Coastal Louisiana Land Changes: 1978-2050*, USGS Open File Report 03- 334 and 39 p. Accessed April 12, 2006, as LCA (Louisiana Coastal Area). 2004. Louisiana - Ecosystem Restoration Study - November 2004 - Volume 4 and Appendix B- Historical and Projected Coastal Louisiana Land Changes: 1978 – 2050, at: <http: //www.lca.gov/final/app_b.htm>

Galloway Jr., G.E., 2003. *America's WETLAND: Campaign to Save Coastal Louisiana*, Report of the Coastal Louisiana Technical Summit (New Orleans, LA. October 16-17 2003). Prepared for the American Society of Civil Engineers Task Committee on America's WETLAND. Accessed on April 12, 2006, at: <http: //www.lacoast.gov/news/press/2004-04-13/report27Jan042.pdf>

Holling, C.S. and Goldberg, M.A., 1971. "Ecology and Planning," *Journal of the American Institute of Planners*, **37(4)**: 221-230.

Izzo, D., 2004 "Reengineering the Mississippi," *Civil Engineering - ASCE*, Vol. 74 No. 7 (July 2004): 39-45. Accessed on April 12, 2006, at: <http: //www.pubs.asce.org/ceonline/ceonline04/0704feat.html>

Penland, S., 2005. "Taming the River to Let In the Sea," *Natural History*, **114(1)**: 42(6).

Rittel, H.W.J. and Webber, M.M., 1973. "Dilemmas in a General Theory of Planning," *Policy Sciences*, **4(2)**: 155-169.

U.S. Army Corps of Engineers, 2004. *Louisiana Coastal Area (LCA) Louisiana - Ecosystem Restoration Study*, November 2004, Volume I: LCA Study, Main Report (New Orleans, 2004). Accessed on April 12, 2006, at: <http: //data.lca.gov/Ivan6/main/main_report_all.pdf>

U.S. Geological Survey, 1995. *Louisiana Coastal Wetlands: A Resource At Risk.* (Marine and Coastal Geology Program, 1995). Accessed on April 12, 2006, at: <http: // marine.usgs.gov / fact-sheets / LAwetlands / lawetlands.html>

Vasishth, A., 2008. "A Scale-Hierarchic Ecosystem Approach to Integrative Ecological Planning," *Progress In Planning*, **70**: 99-132.

15

Non-timber Forest Produce for Livelihood Security

R.K. Pandey & Satvant Kaur Saini

INTRODUCTION

In the recent past, there has been increasing interest in the potential of NTFPs collection and uses of forest resources for achieving livelihood security of poor forest dependent communities. Unfortunately deforestation is taking place at an alarming faster rate all over the world than ever before. It is estimated that nearly 1.8 per cent of the forests are being degraded every year at global level. The major threat to deforestation is considered to be the growing population. The traditional communities living at forest fringes use forest resources in various ways. Understanding of the resource use pattern of such communities provides a basis for assessing the forests dependence of local people and to what extent restrictive biodiversity conservation strategies would affect their livelihoods. In terms of livelihood of the forestry based traditional communities, conservation and sustainable management of utilizable NTFPs are very important both for subsistence and economic upliftment.

Deforestation and Forest Degradation

There are approximately 2.5 billion ha of closed forest and 1.2 billion ha of open woodlands and savannas in the world (Sedjo, 1983). In recent decades, anthropogenic pressures resulting from increasing demand of biological resources have led to high rate of deforestation all over the world.

It is estimated that forests covered one quarter of Earth's entire land area in 1950 but only one fifth in 1980 (Botkin and Simpson, 1990). During 1970-80, the world witnessed alarming rates of deforestation in tropical countries. The annual rates of forest loss at the global level during 1980-90 was about 16.3 m ha. Most of this deforestation has occurred in developing countries. For example, Nepal lost more than half of its forest cover between 1950 and 1980. Land clearing for agriculture is one of the principal causes of deforestation in Nepal. In India also, particularly in tribal belts, considerable forest area has been encroached and cleared for agriculture. Deforestation is the second most important cause of increase in the atmospheric CO_2 concentration, after fossil fuels burning. Besides deforestation, the forests of the world, especially the tropical forests are declining at an alarming rate due to various direct/underlying and natural/anthropogenic causes as shown in Figure 15.1.

In India, forest degradation is greater cause of concern than deforestation. With increase in human population, biotic pressures on forests are continuously increasing. The most alarming problem is that of firewood shortage. Whereas firewood provides only less than 1 per cent of energy in developed countries, it provides one-quarter of the energy in developing countries. Firewood is removed from forests through head loads in quantity much beyond the carrying capacity of the forests. Most of the state governments in India allow free removal of dead, dry and fallen fuel wood from forests. Poor people living inside or near forest areas, collect fuel wood not only for their own consumption but also sell it in nearby towns and it is one of the major source of their livelihood. This practice is leading to over - exploitation of fuel wood in quantity much more than what can be sustainably removed from forests. Another major cause of degradation of forests is over grazing. Over-grazing leads to soil compaction that in turn results into reduction in soil moisture retention and increase in surface water runoff. The young regeneration is not allowed to establish due to excessive grazing pressure. Other forest products are also being over - exploited. Thus, there is a vicious circle of degradation. The overall productivity of Indian forests is already much less than the global average. Human and cattle population dependent on forests is very high. There is a huge gap between the demand and sustainable production limits of forest products. Therefore, there is over-exploitation of forest products leading to further loss of productivity and widening of gap between the demand and supply. These are the major factors also responsible for creating adverse micro environmental conditions for growth of utilizable forest resources particularly, the species of herb, shrub and climber categories.

This degradation is manifest in several ways, some of the indicators are given below (Shukla and Pandey, 2006):

(a) *Soil*: Soil is one of the most valuable natural resources of the Earth but its productivity is going down. Due to excessive collection of leaves, twigs and fuel wood from the forest floor, the whole nutrient cycling is getting disturbed. There is now very little humus and soil organic matter. Over-grazing is leading to soil compaction and reduction in soil porosity. Population of soil micro-organisms is also reducing. In short, there is an overall degradation in physical, chemical and biological properties of soil.

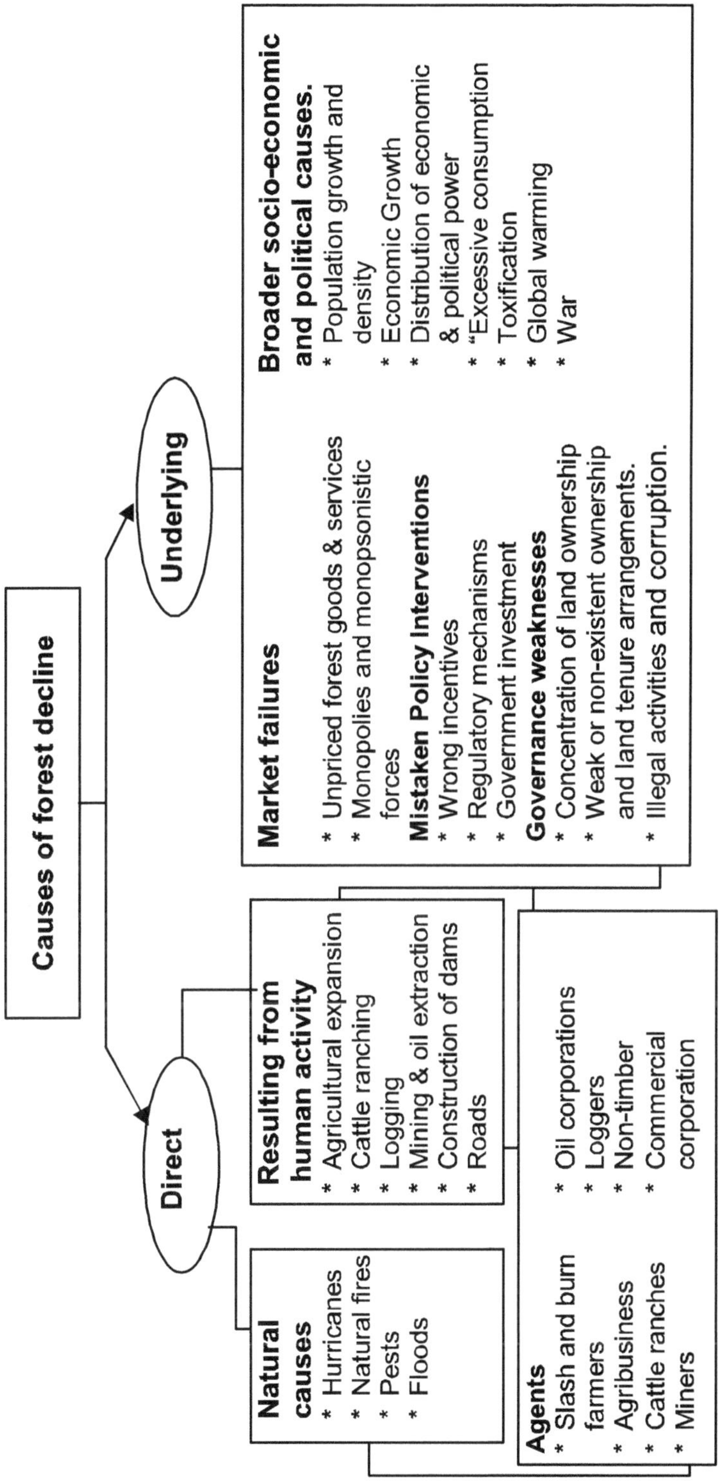

Figure 15.1 The causes of forest decline.

(b) *Water*: Water is another most vital natural resource without which there can be no life on the earth. Due to over grazing and deforestation, the groundwater table is going down and surface run-off is increasing.

(c) *Vegetation*: There is loss of bio-diversity. Many species have already become extinct and many more have reached at the threshold of extinction. The World Conservation Union has calculated that about 12.5 per cent of the world's 2,70,000 species of plants are threatened by forest decline (WCFSD, 1999). Undesirable weeds are replacing useful species. High forests are gradually being converted into coppice forests. Repeated cuttings are resulting into loss in vigour.

We have come across several instances when trees of particular species suddenly start drying from the top and ultimately die. This phenomenon has been witnessed in sal and teak forests of Madhya Pradesh and Chhattisgarh State. However, the exact causes of this drying or dying of trees are still not known but it is obvious that the consequence of overall ecological degradation adversely impact the existence of several valuable NTFPs in natural forest ecosystem.

NTFPs: A Source of Livelihood

A valuation undertaken by Ministry of Environment and Forests, Govt. of India (1995) for the country estimates that 220 million tones of grass and green fodder and 12 million m^3 of timber are removed from India's forests annually. These products are estimated to be worth US$ 10 billion (Mukherjee, 1994).

In India Non-timber forest products are playing major role not only for the subsistence, but also contribute substantial amount in the annual income of rural and tribal people. NTFPs provide about 40 per cent of total official forest revenues and 55 per cent of forest-based employment. Nearly, 50 million people living in and around forests rely on NTFPs as a critical component for their subsistence (World resource Institute, 1990). In Madhya Pradesh, the NTFPs which are primarily collected by tribal women (*i.e.* members of local indigenous groups) are worth more than Rs. 21 billion (US$ 700 million) annually (World watch Institute, 1991). NTFPs also serve as a vital livelihood safety net in times of hardship. Furthermore, in a study conducted for tribal inhabited Baiga chuck area of Madhya Pradesh it was estimated that about more than 60 per cent of annual income of these communities contributed from forests (Shukla and Pandey, 1993). An important feature of the dependence is that almost all NTFPs are harvested from natural forests. Moreover, more than 80 per cent of raw materials required for Indian system of medicines system are still supplied from natural forests. Consequently, extraction of these resources is often intense, exhaustive and unsustainable. There is an urgent need to develop alternative income sources along with adequate management interventions and awareness of consequences of vanishing indigenous forest resources.

Ecological Costs of NTFPs

Tropical forests of the state are potentially rich and supporting enormous range of NTFPs occurring in all natural forest types. More than 700 plant species have been

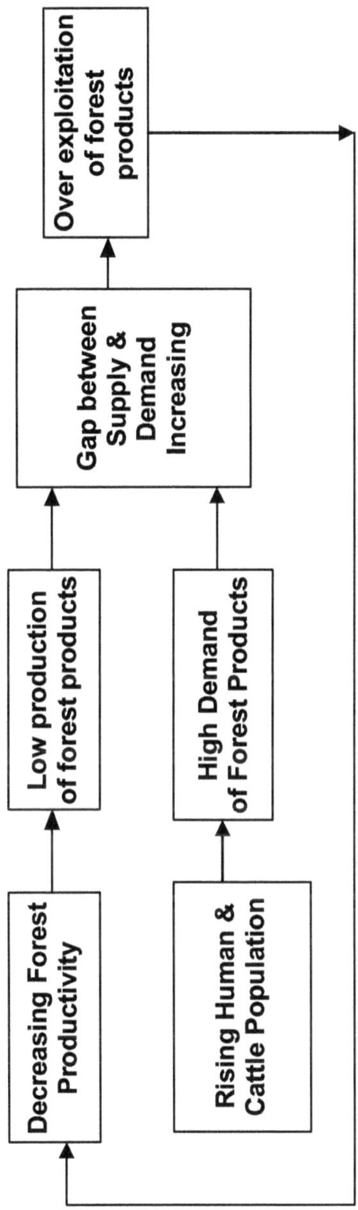

Figure 15.2: Vicious circle of forest degradation.

categorized as potential NTFPs from tropical forests of Madhya Pradesh (Pandey and Shrivastava, 1996). However, the natural resources especially the economically important NTFPs have been plundered without sufficient thought for the long-term consequences of unscientific and non-sustainable exploitations. Prevailing biotic factors such as grazing, fire and illicit felling have caused havoc with the natural forest resources and threatened the existence of various economically viable utilizable resources by creating non-congenial habitat conditions to grow *in-situ*. Several such NTFPs especially the tiny herb, shrubs and climbers are most vulnerable to these biotic pressures (Prasad and Pandey, 1987; Prasad *et.al*, 1990). As a consequence of over extraction of these resources, presumably due to extraction pressure, several species of medicinal and commercially valuable plants are disappearing from the area and are facing adverse environment to survive.

People Centred Approach: Joint Forest Management

Indian National Forest Policy, 1988 has laid down laudable objectives *viz.*, maintenance of environmental stability through preservation and where necessary restoration of ecological balance that has been adversely disturbed by serious depletion of the forest, conserving the National heritage of the country by preserving the remaining natural forest with the vast variety of fauna and flora which represents the remarkable bio-diversity and genetic resources of the country.

Joint Forest Management through community participation, by "Village Forest Protection Committees" (VFCs), "Forest Protection Committees" and "Eco Development Committees" (EDCs) are playing specific role in the overall local natural resource management. However, in the prevailing management system, NTFPs could not get proper attention as far as sustainable management is concerned. Due to lack of basic information and sustainable harvesting pattern in the management system, most of the NTFP species are disappearing from the forest ecosystem. Gradual increase of spacio-temporal pressure on such resources in the absence of adequate management interventions, valuable NTFPs are in the verge of extinction from natural forest ecosystem. Most of the associate species of forest ecosystem belonging to various categories *viz.*, herbs, shrubs, trees, and climbers of annual, biannual and perennial habits are sensitive and struggling for their survival in depleted non congenial habitat conditions. Utilizable plant species need a specific sustainable harvesting management system, which may be species specific on the basis of utilizable parts of the plants in *in-situ* conditions.

Ecosystem Approach for Sustainable Forest Management

Forests constitute a complex, vital and dynamic ecosystem. It provides habitat and sustenance to a range of living organisms, starting from mammals, reptiles birds and insects to plants and soil micro-organisms, survival of each being dependent on others. Besides this, forests provide invaluable environmental services. Thus, it is essential that forests should be managed in a holistic manner by adopting ecosystem approach. Earlier, forests used to be treated as a factory to produce timber for the state and working plans emphasised merely on the sustainable timber yield, ignoring the importance of non-wood forest products and the environmental services provided by the forests. But now, there is a paradigm shift in the attitude of forest managers and other stakeholders and the view that forest should be managed as an ecosystem in totality, is gradually gaining ground.

Users Communities Participation

Rural communities living in and near forests, particularly, the tribal communities, have been traditionally dependent on forests for food, fodder, fuel wood and other requirements and they are the people who have the highest stake in the conservation and sustainable management of forests. Prior to the advent of British rule in India, large tracts of common land, including forest areas, were available and designated for various uses under community forests. The village community collectively guaranteed Common rights. They were also made secure by customs revered as "sacred". The policy of "reservation" of forests adopted by the British rulers in India led to the alienation of people from the forests. This colonial policy of forest management continued even after Independence till eighties.

As a result, forests continued to be subjected to encroachments, illicit felling, rampant forest fires, excessive grazing and unsustainable removal of forest products as participation of local people was not sought in the planning, management, protection or development of forests. It was therefore, realised that the management techniques depending upon the custodial policy had failed to protect the forests in India.

The National Forest Policy of 1988, for the first time, took note of the importance of protecting the livelihood security of the people who live in and around the forests, along with the ecological and environmental objectives of forest management. It was clearly enunciated that the holders of customary rights and concessions in forest areas should be motivated to identify themselves with the protection and development of forests from which they derive benefits. Having regard for the symbiotic relationship between the rural people and forests, the primary task of all agencies responsible for forest management should be to associate the tribal people closely in the protection, regeneration and development of forests as well as to provide gainful employment to the people living in and around the forests. This led to the concept of "eco-development" where 'eco' denotes to both 'ecological' as well as 'economic'. Thus "eco-development" approach paves the way for the ecological development of forests, along with the economic development of the forest dwellers.

Involvement of Forest Dependent Rural Communities as Co-Managers: Joint Forest Management (JFM)

In pursuance of the National Forest Policy of 1988, the Government of India, vide Circular No. 6-21/89.FP dated 1st June 1990, issued guidelines to state governments to organise the beneficiaries (forest dependent rural communities) into institutions for ensuring people's participation in forest regeneration and protection. Such beneficiaries would be entitled to the usufruct of all non-wood forest products and a share in proceeds of final tree harvest, subject to successful protection. The (FPCs) are constituted in sensitive forest areas for the protection of 'dense' forests (forests having more than 40 per cent of canopy cover) VFCs are constituted in open forest areas (forest having less than 40 per cent canopy cover) and the EDCs in the buffer zones of protected areas *i.e.* national parks and wildlife sanctuaries. Madhya Pradesh is the leading state in constituting JFM committees. More than 14,000 such committees have been formed by involving 17 lakhs rural forest dependent families to manage 6000 km^2 of forest land, which is 63 per cent of the total forest area of the State.(Table 15.1).

Table 15.1: JFM programme in Madhya Pradesh

Sl.No.	*Name of the Committees*	*No. of Committees*
1	Village Forest Committees (VFCs)	9035
2	Forest Protection Committees (FPCs)	4201
3	Eco-development Committees (EDCs)	785
	Total	14021
	Total Forest Area under JFM	60.000 sq. km. (63 per cent of total forest area of the State)
	Total no. of Families involved in JFM	17 lakhs

Source: M.P. Forest Department (www.mp.nic.in/forests).

The Joint Forest Management Programme although gaining monmentum but will take its own time to come in mainstream of joint management system. It is now widely accepted that the forests can be conserved and developed only through collaborative efforts with the dependent people. JFM is gradually establishing roots in the forestry system and becomeing more open and responsive to the needs and aspirations of the people.

Self-Assessment Technique for SFM through Users Communities

There must be a comprehensive planning system for assessment and management of utilizable forest resources at community level. A Self Assessment system (Pandey and Saini, 2001) seems to be an innovative method for assessing status and sustainable management of the forest resources *in-situ* at users community level in the tropical forests in developing countries like India.

In one of the case studies, an innovative "**Self-Assessment**" methodology adopted for assessment of status of utilizable resources in *in-situ* through users community participation has shown encouraging and positive impacts in sustainable harvesting and management of NTFPs in study sites. Assessment of NTFPs through people's participation in JFM areas essentially has two components *viz.* (i) assessment of productivity and (ii) assessment of current level of harvesting. By this means species-specific information of utilizable NTFPs could be assessed and monitored in a regular way through direct involvement of local people.

Competent and properly motivated work force has to be developed which should not only to list the recommended practices, but also to explain the conditions under which they apply and the reasons considered important.

In the comprehensive planning system,the following points need to be considered for development of strategies for forest resource management through people's participation:

- ☆ Identification of potentially rich NTFPs areas in natural forests
- ☆ Resource inventory
- ☆ Present status/availability of NTFPs (quantitative assessment) and their harvesting regimes
- ☆ Consequences of over use of resources on sustainability

At community level, technology should be simple, energy efficient (economically viable) and environmental friendly.

Eco-Sustainability

The term 'eco-sustainability' refers to both the ecological sustainability of forests as well as the economical sustainability of dependent rural communities. Both are complimentary to each other. Increasing biotic pressures of the burgeoning human population in developing countries of this region, has made the management of forests and other natural resources unsustainable as these resources are being over exploited. Poor rural population residing in and around forests has to depend mainly on forests for their requirements of fuel and fodder. Change in consumption pattern of these forest products is possible only with the economic upliftment of the dependent rural population. It has also been observed that people participated enthusiastically in JFM where village resource development works and the forest department and other developmental agencies undertook income and employment generating activities. It is now evident that conservation and sustainable management of forests is possible only through active participation of dependent rural communities and that in turn, is dependent on simultaneous economic development of villages.

From the above discussion, it follows that sustainable economic development is essential for the success of participatory forest management. Releasing the importance of economic development, most of the state governments initiated entry point development activities while launching JFM programmes to win the confidence of local people. However, these development activities require sustained funding which is not possible unless some self – sustained mechanism is evolved to generate enough funds for economic development works as a bye - product of sustainable forest management itself. It has been observed during the implementation of JFM programmes in many states that JFM programmes were launched, accompanied with economic development activities under some externally funded forestry projects. Villagers also displayed lot of interest in the protection and development of forests under JFM, as there was a massive flow of funds for the economic development of villages also. But with the winding up of the externally funded projects, the funding source dried up and it was difficult to sustain the interest of villagers in JFM. More difficulty was faced in the expansion of JFM programme in new villages in the absence of entry-point development activities.

Sustainable Forest Development and CDM

Clean Development Mechanism (CDM) under the Kyoto Protocol provides a potential source of funding to developing countries for projects of sustainable management. Article 12.2 of the Kyoto Protocol specifies that CDM projects should assist host countries in achieving sustainable development.

According to Pearce *et al* (1988) and Bass *et al* (2000), sustainable development can be viewed as the creation of opportunities and capacities for raising per capita well-being, *i.e.* the quality of life. Carney (1998) has specified the following five types of assets (or capital) that need to be expanded or made more productive to achieve this end:

1. Natural capital, such as land, forests and water
2. Physical assets, such as farm tools, roads, electricity etc.
3. Financial capital, such as savings
4. Human capital, such as health and education, and
5. Social capital, such as membership in FPCs/VFCs etc.

The concept of sustainable development includes equity, which can be achieved by increasing the well-being of less advantaged populations. Thus, sustainable development not only promotes opportunities for increasing the above five forms of capital but also facilitates empowerment, *i.e.*, strengthening the participation of poor people in local decision making (World Commission on Environment and Development, 1987)

Natural forests and plantations represent more than natural capital. Income from the sale of forest products such as *tendu* leaves, medicinal plants and gums etc. provide financial capital. Forest incomes are a vital economic buffer, particularly for women, children and the poorest households in village communities (Byron and Arnold, 1999). Rehabilitation of degraded forests and eco-development activities under JFM programmes help in the creation of physical assets and human capital. JFM has proved to be a powerful tool for enhancing the social capital also by enlisting the people's participation in forest management. Thus, JFM leads to sustainable development and qualifies for CDM.

Forest-Carbon Projects and Livelihood Benefits

Different types of forest carbon projects like forest rehabilitation and regeneration, strict protection of forest areas, multiple-use community forestry (Timber, NTFPs fodder etc.), agro-forestry, pulp wood or timber plantation etc. have potential to mitigate climate change through two major strategies *viz.*, afforestation and reforestation (AR) and averted deforestation (AD). AR helps to mitigate climate change by establishing additional forest area that sequesters carbon as trees grow and then stores carbon for the life of the standing trees. The second strategy AD protects standing forest carbon stocks that would otherwise be lost. While the focus has so far been on strictly protected areas under AD strategy, it is now being realized that multiple-use community forestry can provide greater livelihood benefits. However, AD projects are not presently eligible under CDM (Smith and Sherr, 2002).

In regions where natural forests have been depleted due to intense biotic pressure, the dependent communities can be involved in forest rehabilitation and protection. Techniques such as assisted natural regeneration (ANR) and coppice management can accelerate forest recovery and improve quality (Saxena, 1997)

Participatory Carbon Management

JFM had already occupied more than 7 million ha of forest and in India by 1999 (MoEF, 1999) and it has increased to a great extent between 1999 and 2003. FSI report suggests that after JFM coming into operation, the forest cover in India has almost stabilized (Rangachari and Mukherji, 2000). It is obvious that JFM has been extremely successful and it has now come to stay as a viable strategy of sustainable forest management. The key stone for the success of JFM is participatory approach by involving the dependent rural communities, who are the greatest stake holders, in the protection and management of forests. However, as already stated, the interest of these rural communities can be sustained if village development works are also carried out simultaneously.

Now, the question arises from where can we get the funds required for village development sustainable. Emerging global carbon markets can be the answer. Studies carried out in central India revealed that teak and dry deciduous forests under community protection sequester 1 to 3 metric tons of carbon per ha per day and that the best opportunities were in degraded forests with crown cover under 40 per cent (Poffenberger *et al.*, 2001). According to the land-use/land-cover statistics of India generated by remote sensing techniques, the areas under non-forest degraded lands are currently 93.68 million ha and 35.89 million ha (Singh and Lal, 2000). They have estimated that if a conservative managed forest productivity of 5.5 tonnes per annum could be attained on only 40 million hectares of the available surplus degraded land in India through managed forestry programmes such as short rotation forestry plantations, it is possible to obtain carbon mitigation of about 3.32 Gt in the next 50 years, with an annual reduction of about 0.072 Gt of carbon.

It is therefore, proposed that a 50-year contract be entered into by the village level JFM committees (FPCs/VFCs/EDCs) with the outside financing agencies. The financing organizations would agree to finance carbon sequestered at a fixed rate of US$ 10 per ton per year during the first 25 years of operation, after which the rate would increase to US$ 20 per ton per year (Poffenberger *et.al.*2001). At this rate, there is a possibility of generation of funds to the tune of about 720 billion to 1440 billion US$ annually, which is a gigantic sum to sustain JFM programme for the eco-sustainability of forests and dependent rural communities. Economic upliftment programmes through alternative means

Acknowledgements

We acknowledge the financial assistance received from Council of Scientific and Industrial Research (CSIR) GoI, New Delhi. Authors are also thankful to the Director, SFRI for encouragement and support for successful completion of the work.

References

Bass, S., Dubois, O., Moura-costa, P., Pinard, M., Tipper R. and Wilson, C., 2000. Rural livelihood and carbon management. IIED Natural Resource Issues Paper No.1. International Institute for Environment and Development, London.

Botkin, D.B. and Simpson, L., 1990. The first statistically valid estimate of biomass for a large region. *Biogeochemistry*, **9**: 161-174.

Byron, N. and Arnold, M., 1999. What future for the people of the tropical forests? *World Development*, **27**: 789-805.

Carney, D., 1986. Implementing the sustainable rural livelihood approach In: Carney, D. (ed.) *Sustainable Rural Livelihoods: What Contributions can we make*? 3-23, DFID, London.

MoEF (Ministry of Environment and Forests), 1999. National Forestry Action Programme in India. Government of India, New Delhi.

Mukherjee, A.K., 1994. India's forests: a status report: concepts, definitions, trends, controversies. Paper presented at the International Workshop on India's Forests Management and Ecological Revival. New Delhi, 10-12 February.

Pandey, R.K. and Shrivastava, J.L., 1996. A material for forest flora of Madhya Pradesh State Forest Research Institute, Jabalpur, M.P. S.F.R.I. Bulletin No.28.

Pandey R.K. and Saini, S.K., 2001. Pachmarhi Biosphere Reserve ke buffer zone mein satat van-prabandhan ke adhar evum suchkank (Criteria and Indicators) ka vikas I : Ikai star per jan samuday kee bhumika (In Hindi). *Vaniki Sandesh*, **25(1)**.

Pearce, D., Bay, B., Newcombe,J., Brunnelo, T. and Bello, T., 1988. The Clean Development Mechanism, benefits of the CDM for developing countries. Unpublished report. CSERGE, University college, London.

Poffenberger, M. (ed) Ravindranath, N.H.: Pandey, D.N., Murthy, I.K., Bist, R. and Jain, D., 2001. Communities and climate change: The Clean Development Mechanism and village based forest restoration in Central India. A case study from Harda Forest Division, Madhya Pradesh, India. Community Forestry International, Inc. and the Indian Institute of Forest Management, Bhopal: Santa Barbara, California, U.S.A.

Prasad, Ram and Pandey, R.K., 1987. Medicinal wealth of central India: I Potential of Indigenous medicinal plants of Eastern Madhya Pradesh. *J. Trop. For.*, **2(4)**: 287-297.

Prasad, Ram, Pandey, R.K. and Bajpai, A., 1990. Ethno-medico-botanical studies on indigenous medicinal plants of Lamni and Achanakmar Forests of Bilaspur District of Madhya Pradesh. Proceedings of National Seminar on Medicinal and Aromatic Plants held at S.F.R.I. Jabalpur on 6-7 April. 1990.

Rangachari, C.S. and Mukherji, S.D., 2000. *Old roots new shoots. A study of joint forest management in Andhra Pradesh*, India Winrock International and Ford Foundation, New Delhi.

Saxena, N.C., 1997. *The Saga of Participatory Forest Management in India.* Special Publication. CIFOR, Bogor, Indonesia.

Sedjo, Roger A., 1983. *The Comparative Economics of Plantation Forestry: A Global Assessment, Resources for the future.* Research paper. John Hopkins Univ. Press. Baltimore, U.S.A.

Shukla, P.K. and Pandey, R.K., 2006. Participatory carbon management for eco-sustainability of forests and dependent rural communities. *Advancing Frontiers of Ecological Research in India,* (Eds.) A.K. Kandya and Asha Gupta. Published by Bishen Singh Mahendra Pal Singh, Dehradun, India, pp. 125-141.

Shukla, P.K. and Pandey, R.K., 1993. Tribal life and forests : A case study of selected forest villages in Dindori, Tahsil of Mandla District in Madhya Pradesh. *J. Trop. For.,* **9(4)**.

Singh, Roma and Lal, M., 2000. Sustainable forestry in India for carbon mitigation. *Current Science,* **78(5)**: 563-565.

Smith, J. and Scherr, S.J., 2002. Forest carbon and local livelihoods. CIFOR Occasional Paper No.37.

World Commission on Environment and Development, 1987. Our common future. Oxford University Press, Oxford, U.K.

World Commission on Forests and Sustainable Development (WCFSD), 1999. Our forests our future. Summer Report of WCFSD, Canada; Winnipeg.

World Resource Institute, 1990. *The World Bank in the Forest Sector: A Global Policy Paper.* Wasteland News, **8(2)** : 6-12.

World Watch Institute, 1991. Quoted in Wasteland News (July-August).

16

Conservation and Management of Bhopal Lakes

Sanjeev Sachdev

INTRODUCTION

The construction of storage reservoirs is an age old practice in India. Former rulers have contributed significantly by constructing large number of impoundments for providing drinking water to the people in their capitals and elsewhere. This was particularly necessary in arid, semi arid and other regions with highly erratic rainfall. Upper lake of Bhopal, arguably the oldest among the large man-made lakes in central part of India, falls under this category. Constructing an earthen dam across the Kolans, a rainfed tributary of Betwa river, this Lake was created in the early 11th century AD by the King Bhoj, the ruler of Dhar. Its large spread made it so famous that all other lakes are considered simply 'ponds' in its comparison.

In the late 18th century AD, another lake on the downstream of earthen dam of Upper lake was constructed by Nawab Chhote Khan, the Minister of Nawab Hayat Mohammed Khan to enhance the beautify of the city.

With the passage of time, the administration of Bhopal City changed hands several times, and in the year 1956 it became the capital of State of Madhya Pradesh, the then largest state of India. Since then it noticed tremendous influx of people and consequent urban development especially on the northeast fringe of the Upper Lake, as well as around the Lower Lake (Figure 16.1). This caused increase in demand for potable water and thus pressure on the Upper lake. Consequently the storage capacity of the Upper lake had to be increased through raising the height of spillway. Though

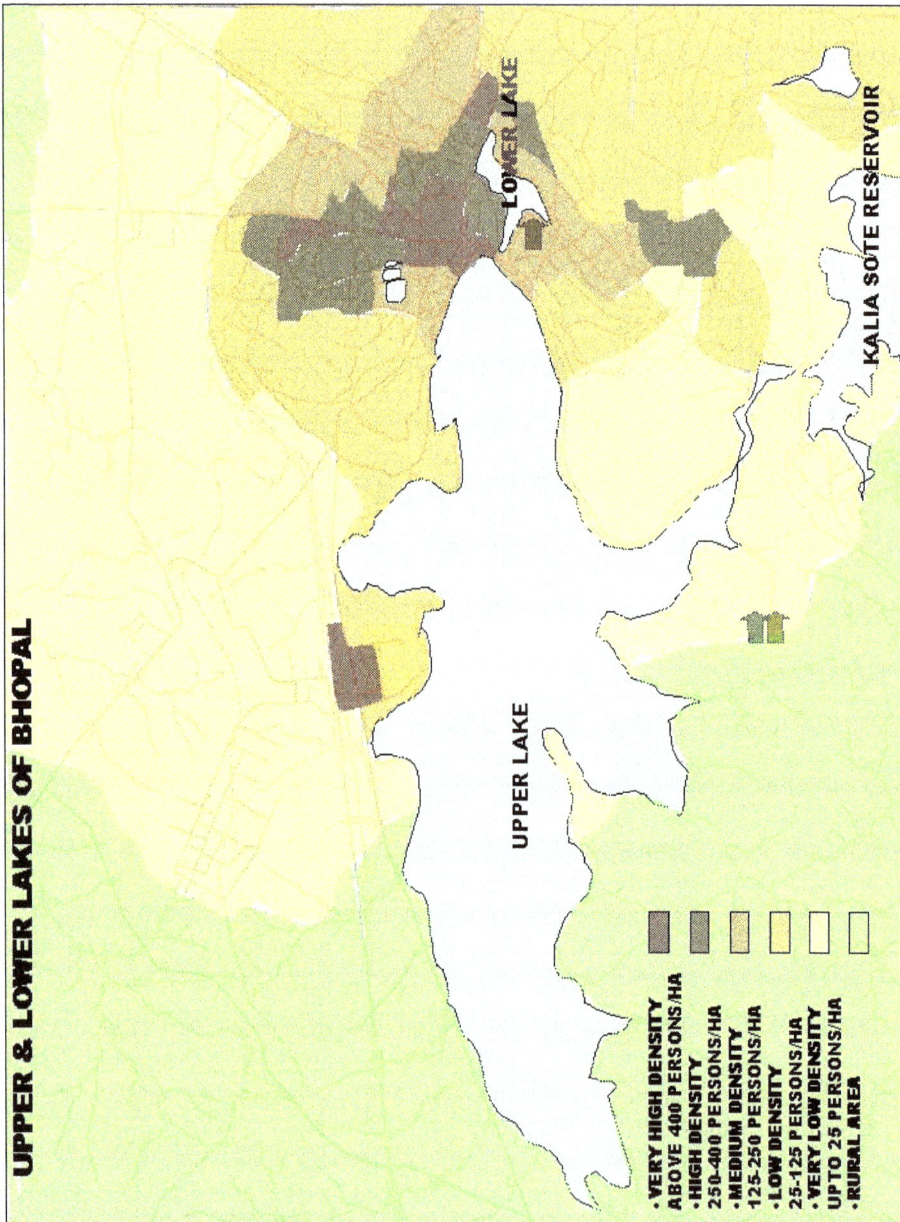

Figure 16.1: Upper and lower lakes of Bhopal.

this had helped in easing the water supply, the increased anthropogenic activities in the catchment caused increased inflow of silt, untreated sewage, nutrients and pesticides from urban and rural areas and thus deterioration of water quality of the lakes. However, in view of its ecological importance, the Ministry of Environment and Forests, Govt. of India has recognized the Upper and Lower lakes as wetland of national importance and designated them as Bhoj Wetland and in 2002, it was declared a Ramsar site. An integrated plan for the conservation and management of the lakes was conceived, and implementation of the same was started in 1995 through financial assistance of Japan Bank for International Corporation of Japan.

The Bhopal Lakes (Bhoj Wetland)

The Bhoj Wetland, a Ramsar site, comprise of Upper and Lower Lakes. The salient features of the lakes are as under:

Sl.No.	*Features*	*Upper Lake*	*Lower Lake*
1.	Constructed in	11th Century A.D.	Late 18th Century
2.	Type of dam	Earthen	Earthen
3.	Location:LatitudeLongitude	23°12' - 23°16' N 77°18' - 77°23' E	23°14' - 23°16' N 77°24' - 77°25' E
4.	Catchment area (Sq.km.)	361	9.6
5.	Submergence area at FTL (Sq.km.)	31.00	1.29
6.	Full tank level (MSL) (m)	508.65	499.88
7.	Dead storage level (MSL) (m)	503.53	499.88
8.	Storage capacity (M Cum.)	117.05	4.3
9.	Maximum depth (m)	11.7	9.4
10.	Source of water	Rain water	Seepage from Upper Lake and Domestic Sewage
11.	Main use of water	Potable water supply and Recreational	Washing and Boating
12.	Inflow points (Nos.)	31	28
13.	Sewage water inflow (MLD)	50.47	31.63

Catchment Area

The catchment area of both the lakes displays a complete range of urban and rural activity with varying intensities. While the catchment of Upper Lake spreads partly within the city limit of Bhopal and extends south west wards up to adjoining district of Sehore, the catchment area of Lower lake is located within the city limit. The break up of the land-use pattern of the catchment of Upper Lake is as follows:

1.	Built up area	20.9 (Sq.km.)
2.	Crop land	213.1 (Sq.km.)
3.	Plantation	9.6 (Sq.km.)

4.	Open Forest	5.2 (Sq.km.)
5.	Land with scrub or without scrub	87.7 (Sq.km.)
6.	Barren rocky/stony	8.4 (Sq.km.)
7.	Other Lakes/Ponds	16.1 (Sq.km.)
	Total Catchment Area	**361.0 (Sq.km.)**

While major part of Lakebed and the catchment of Upper lake has black cotton soil, the lake bed and the catchment of Lower lake is rocky.

Importance of Bhoj Wetland

The Upper Lake in the capital city region of Madhya Pradesh - Bhopal, along with the Lower Lake are collectively termed as the Bhoj Wetlands. A Ramsar site, the waterbodies are owned by the Bhopal Municipal Corporation, and are the collective heritage of the entire city. The larger Upper Lake has many direct and indirect uses. The lake is (1) a major source of potable water, (2) a recreational hub – with boating, a cruise ship, walkways, restaurants, (3) a tourist destination, (4) a bird habitat and a Ramsar Site, (5) a source of livelihood (fishing, boating, chestnut farming etc), (6) a cultural landmark and (7) a micro-climatic influence on the city. It is also (7) a sink for some of the city sewage, although this use is reducing with enhancement and use of sewage treatment plants.

Balancing these multiple uses is a challenging task especially as many uses are competing - *e.g.* recreation and livelihoods impact drinking water quality and bird habitat. Further, operational responsibilities for permitting and managing different uses rest with multiple agencies. Of these uses drinking water supply is currently perhaps the most important function of the lake and improved quality of raw water to Bhopal is an important objective for the city. In addition, the Upper Lake's importance as a habitat for hundreds of migratory bird species, and reflected in its Ramsar nomination, is an important biodiversity value of the lake. The importance of Bhoj Wetland is summarized below:

- ✰ While upper lake is the principal source of drinking water (40 per cent) for the city of Bhopal, the Lower Lake meets out the requirement of raw water and enhances the beauty of the city.
- ✰ The Upper Lake is rich in bio-diversity, principal components being phytoplankton, zooplankton, macrophytes, aquatic insects and avifauna (both resident and migratory). The details of bio-diversity of the lake is as follows:

1.	Macrophytes	106 species (Belonging to 87 genera of 46 families), including 14 rare species.
2.	Phytoplankton	208 species comprising of 106 species of Chlorophyceae, 37 species of Cyanophyceae, 34 species of Euglenophyceae, 27 species of Bacilariophyceae and 4 species of Dinophyceae
3.	Zooplankton	105 species (Rotifera 41 + Protozoa 10 + Cladocera 14 + Copepoda 5, Ostracoda 9, Coleoptera 11, Diptera 25)

4.	Fish fauna	43 species (natural and cultured species)
5.	Avifauna	180 species (Migratory-52,Local migratory-29and Local-99)
6.	Insects	98 species of 10 families.
7.	Reptiles and Amphibians	>10 species (including 5 species of tortoise)

Further a buffer zone along the fringe of the wetland has been created through plantation of 51 species of angiosperms with adequate representation of medicinal plants.

Problems and Causes

The Upper Lake is an elongated water body with a water submergence length of 12 km and maximum width of 1.75 km at Full Tank Level (FTL). The submergence v/s catchment area ratio of about 1:10 provides ample opportunity for Lake Recharge; the lake thus attains FTL with normal rains. The catchment of the upper lake may be categorized into three broad sectors. The relative area of these three sectors is:

- ✰ Urban Catchment 20 per cent
- ✰ Semi-urban catchment 10 per cent
- ✰ Rural Catchment 70 per cent

The type of pollution received from each sector is as follows:

Urban Sector	*Semi-Urban Sector*	*Rural Sector*
Untreated Sewage Inflow Municipal Solid waste dumping Human Intervention Anthropogenic activities Bathing, washing of clothes and vehicles	Partly urban and partly rural waste joining to lake.	Agriculture waste Chemical Fertilizer Residue Cattle Intervention/grazing Inflow of silt/soil Municipal Solid waste of Villages Pesticides and Insecticides

Due to tremendous population growth of the city (about 70,000 in 1951 to about 13 lakhs in 2001) and rapid urban development around Lower lake and on the eastern and northern fringes of Upper lake especially during 2nd half of the last century subjected both the lakes to various environmental problems resulting in deterioration of their water quality. The major causes of environmental problems of the lakes are as under:

Problems	*Causes*
1. Reduction of storage capacity of lakes	i. Inflow of monsoon runoff and dry weather flow carrying silt and organic materials from urban and rural catchments.
	ii. Addition of clay and bio and non-biodegradable materials through Idol immersion.

	Problems		*Causes*
2.	Obstruction to smooth flow through the spill channel of Upper lake		Deposition of silt
3.	Deterioration of water quality	i.	Inflow of untreated sewage.
		ii.	Dumping of Municipal wastes not collected by the Municipal Corporation.
		iii.	Addition of organic and inorganic materials through Idol and Tazia immersion.
		iv.	Direct human intervention and encroachment of fringe areas.
		v.	Addition of detergents used for washing clothes.
4.	Flourishing growth of invasive aquatic plants and erosion of biodiversity.		Nutrient enrichment of lakes due to inflow of sewage and agricultural wastes.
5.	Erosion of Takia Island of religious importance causing siltation within Upper lake.		Thrust of waves.
6.	Threat to stability of dam of Upper lake.		Hindrance in flood discharge due to silt deposition within spill channel of Upper lake.
7.	Threat to stability of spillway structure.		Increase in traffic pressure on the bridge over the spillway due to expansion of city.

Interventions to Improve Water Quality of Lakes as well as for Sustainability of the City

An integrated conservation and management programme (Bhoj Wetland Project) having both preventive and curative measures are being implemented since 1995. The action plan (Figure 16.2) is being implemented through 19 subproject action programme, most of which have been completed benefiting the city people. Measures implemented include the following:

Demarcation of No Construction Zone

The Bhopal Development Plan 2005 prohibits construction within 50m of the Full Tank Level (FTL) of Upper lake. Accordingly, a 50m wide strip of land all along the FTL of Upper lake was demarcated and a physical barrier along Lower lake was constructed to prevent human intervention.

Creation of Buffer Zones between the Lake and the Human Settlements

i) Encroachments from the Shore of Upper Lake were removed and a 5.4 km Link road on the north-east and a 2.5 km long Lake View Promenade on the south east fringe of Upper Lake were constructed.

ii) *Buffer Zone Plantations*: In order to prevent encroachment for human settlements and cultivation and grazing within the lake area, buffer zones have been created particularly in the Western, Southern and Northern fringe of Upper lake. Besides this, intensive plantation has been carried out in the

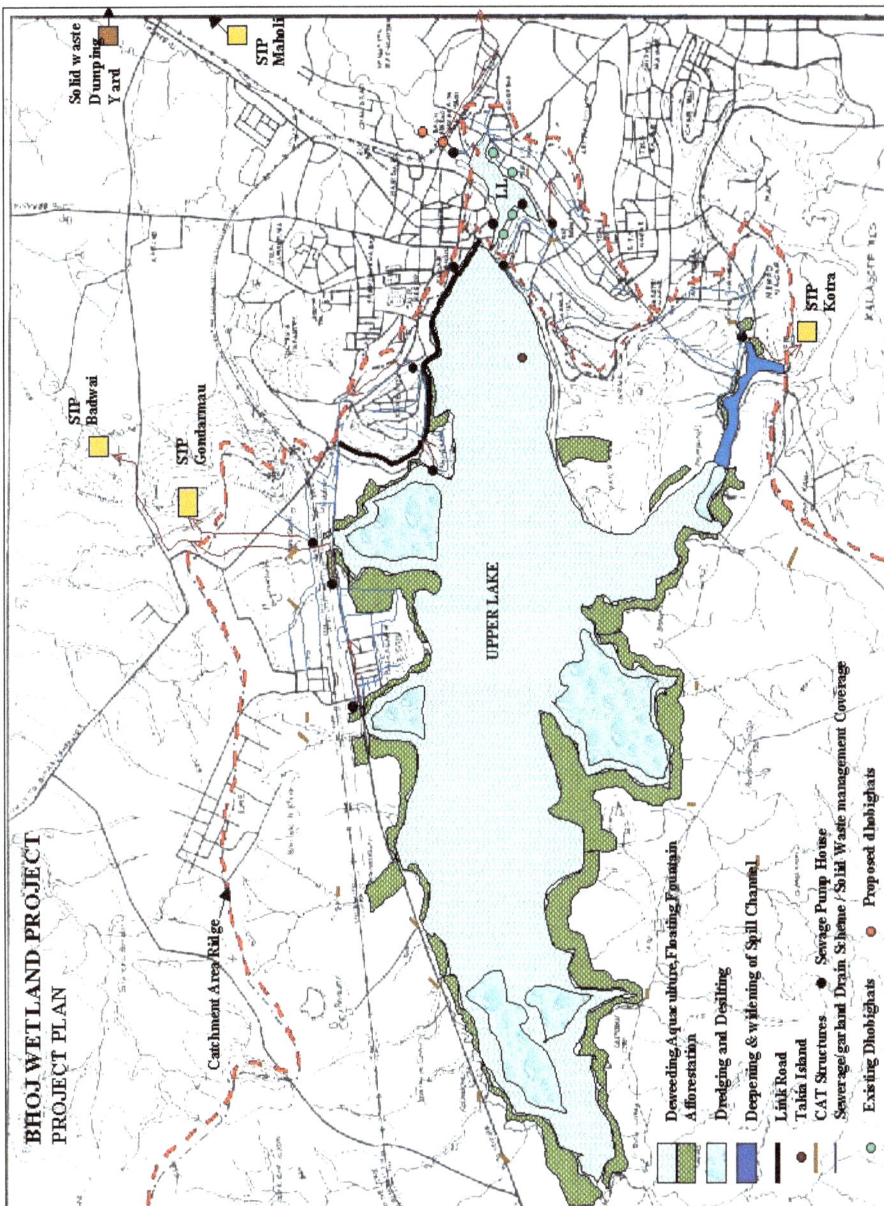

Figure 16.2: Bhoj wetland project plan.

catchment area of the lake to control soil erosion. The species selected are either Bio-mass producing or having medicinal properties and are tolerant to both flooding and drought conditions. About 17 lakhs plants have been planted in over 1000 hectare land for over a period of 12 years. In the programme under social forestry, about 2.06 lakhs of plants have been distributed to the farmers and were planted in the catchment area.

Catchment Area Treatment

In order to mitigate inflow of silt, agricultural residues and other wastes into the lakes, 75 check dams made of loose boulder/gabion structures and 2 silt traps having a cumulative silt trapping capacity of 3.64 lacs cum have been constructed across 31 inlet channels.

Sewerage System

Infrastructure (laying of 86.7 km pipeline through congested human settlements and construction of 8 sewage pump houses and 4 treatment plants) for diversion and treatment of 56 MLD domestic sewage is being developed.

Improvement of Solid Waste Management System

Infrastructure of Bhopal Municipal Corporation was strengthened. Outcome being additional collection and disposal of 70 MT of solid waste from the 18 municipal wards located in the urban catchment of the lakes

Prevention of Pollution Due to Washing Activities

Development of rehabilitation sites for 250 washerman families outside the catchment of Lower lake having all facilities for living and washing has been completed. Development of vacated land as buffer zone having gardens and parks is likely to be completed by Oct. 2003.

Excavation of Silt

Desilting and dredging was carried out to remove volumes of silt, sediments and nutrients that have deposited in the lakes over long periods, resulting in pollution of lake water and luxurious growth of aquatic weeds, algae and bio-organisms promoted by the accumulation of nutrients and reduction in the storage capacity of the lake. Where nallahs and rivulets meet the lakes, silt has accumulated generally at the peripheral area causing the lakes to become shallow and also result in growth of weeds. With accumulation of silt, land formation starts into the lake. Dredging operation was carried out in three placed in Lower lake and about 85,000 cubic meter silt was removed. However, looking to the large extent, technical difficulty in dredging in shallow water depth and in a potable water source near the intake, non-availability of dumping yard nearby, the silt removal in Upper lake was undertaken through dry excavation after making earthen coffer dams and dewatering the area through pumping where required. 2.70 million cubic meter silt was removed from 6 zones around the periphery of the lakes, which is equivalent to offloading of estimated 162 ton of nitrogen and 81 tons of phosphorous from the system.

Deepening and Widening of the Spill Channel

The spillway of the Upper Lake is located at Bhadbhada on the Kaliasote river. Continued deposition of silt during the past 33 years, the existing narrow channel section that had been originally provided has practically obstructed the smooth flow in the 4.41 km long approach channel to the spillway. Further the deposition of silt accumulated near weir over a long period caused high silt pressure on the weir. Insufficient discharge capacity of the spillway system also caused heading up of water in the Upper lake during floods and hence causes a threat to the earthen dam at Kamla Park. The deepening and widening works have been carried out in 2.6 km of the spill channel to accommodate a discharge of 566 cum/sec. and about 9.87 lakh cubic meter of silt was removed. The channel has been given a proper shape to attain the required discharge. In this process also about 67 ton nitrogen and about 34 ton phosphorus was off loaded from the system.

Control of Idol Immersion Activities

Alternate idol immersion site on the spillway of Upper lake was developed and through constant persuasion and intensive awareness campaign, total stoppage of immersion at the traditional site, which was located near the potable water intake point, was brought about. Total diversion of idol immersion from traditional site was achieved in year 2002.

Restoration of Takia Island

The Island in Upper lake has a Mazar of Shah Ali Shah Rahmatullah Aliah which is of religious importance. Over the years, due to wave action its periphery was getting eroded from western side, posing threat to the existence of the Mazar, while on the eastern side the lakes was getting silted. As a remedial measure two rows of toe walls was constructed around the island and filled in between the walls with soil excavated from the silted lakebed.

Weed Removal

Nutrient enrichment of lakes due to inflow of untreated sewage and run off containing organic wastes from urban and rural areas caused excessive growth of aquatic vegetation within the lake area. With a view to offload nutrients and prevent accelerated evapotranspiration of lake water, controlled weed removal operation in about 90 per cent of the submergence area of Upper lake and almost entire area of Lower lake was carried out. Different types of weed such as shoreline (Ipomea fistulosa), emergent (Scirpus roylie and Cyprus rotandus, Polygonum glabrum and Ipomoea aquatica), floating weed (Water hyacinth) and an assemblage of submerged weeds were removed in and around the Upper and Lower lakes starting from January 1999.

Installation of Lake Water Oxygenation Systems

First time application of ozone for improvement of water quality of Lower lake has been done. However, for Upper Lake (a potable water source) only fountains have been used. Thus a total of 15 aeration units (1 ozonizer, 1 ozonizer cum fountain and 4 fountains in Lower Lake and 9 fountains in Upper lake) were installed.

Aquaculture

Herbivore grass carps along with Indian Major Carps were introduced into the lakes to control submerged weeds as well as to maintain ecological balance.

Water Quality Monitoring

A dedicated laboratory has been established to analyze 120 parameters on a regular basis and to assess the impact of preventive and curative measures being taken to improve water quality of lakes.

Construction of High Level Bridge Across Bhadbhada Spill Channel

In view of development on the south-eastern part of Upper lake and to reduce traffic pressure over the age old Bhadbhada bridge cum spillway, a 4 lane bridge across Bhadbhada spill channel has been included in the project. The work is expected to be completed by July 2003.

Public Awareness Campaign

There are involvement of about 50 NGOs, 10 colleges, 25 local schools and the general public in creating awareness among the people of Bhopal to save the lakes from deterioration and the campaign is being continued on a planned manner since 1998.

Interpretation Center

To create awareness among the people of all walks of life, an Interpretation Centre depicting the origin of Bhoj Wetland, ecosystem structure and function, conservation principles, Bhoj Wetland Project activities, future course of action for the sustainable use of lake eco-system are under implementation at the shore of Upper lake which is frequented by the people.

Control of Seepage through the Earthen Dam of the Upper Lake

The 125m wide earthen dam of Upper Lake at Kamla Park was built in the 11th century AD with boulder and morrum filling sandwiched between two stone masonry walls at upstream and downstream faces. Later a lime mortar stone masonry tunnel (1.2m wide and 2.9m high) was constructed about 200 years back. This tunnel fitted with a sluice gate is used for releasing seepage/leakage and raw for the Bhopal Municipal Corporation's water treatment Plant. However, due to improper maintenance and growth of vegetation over the bund the displacement of stone masonry wall on the upstream face of the dam took place. There was deterioration of the tunnel. Consequently, seepage through the dam had increased substantially. Moreover, there was a perceived threat to the stability of the dam itself. Reportedly, the leakage had increased to an alarming situation in-spite of corrective measures attempted by the BMC. Considering the distressed status of the dam, the work of restoration and the tunnel and remedial measures to control seepage was done under the project.

Stakeholder Participation

The participation of ordinary people in the project implementation is an essential and important feature. This has been achieved through a well coordinated awareness programme involving political and religious leaders, District/City Administration, local people, NGOs, students, etc. Following were achieved through stakeholder participation:

i) Diversion of idol and Tazia immersion activities from their traditional sites.

ii) Rehabilitation of washer-men through negotiation.

iii) Development of fisheries and control of aquatic weeds through aquaculture and participation of fishermen.

iv) Afforestation of catchment area through participation of farmers.

v) Reduction of use of inorganic fertilizers due to promotion of organic farming in the catchment through participation of farmers.

vi) Rationalisation of water withdrawal from the Upper lake for potable purposes by the Local bodies.

Impact of Project Interventions

i) Due to construction of silt traps across small feeding channels and creation of buffer zone plantation around the lake sedimentation of the lake was considerably reduced. The silt traps were also found to be very effective in trapping the organic debries from the rural watershed.

ii) Before creation of buffer zone plantation, land was either used for agriculture or infested with shoreline weeds. Either of the situations was not desirable. After plantation and protection, not only general ambience of the area improved but due to appearance of ground flora soil erosion from the area become negligible. The area also becomes a major attraction for the avifauna. An avifauna census completed in the year 2001 has revealed 179 species, which include 52 migratory, 28 local migratory and 99 local avifauna species. Some of them are also arboreal reflecting positive impact of plantation on biodiversity. South-western region of Upper lake became a winter habitat for *Grus antigone*, a vulnerable species.

iii) In the plantation area, the understorey grass species become a regular source of feed for the cattles (stall feeding) of nearby villages. The plantation and protection activities also provided ample job opportunities for the villagers, of whom a significant number were of women. For the development of plant nurseries local women were involved in a big way.

iv) The project implementation has provided many job opportunities for the local people in other areas.

v) Socio-economic conditions of fishermen were improved due to increase in fish production due to planned aquaculture.

vi) The Link road not only becomes a barrier against encroachment of Lake Area, but it helped in easing traffic pressure through the old city.

vii) Idol immersion in traditional site was stopped, and at the new site it became possible for the local people to use the materials used for idol immersion and thus adverse impact as water quality was reduced.

viii) Due to reduction of inflow of solid waste and silt, operation of water oxygenation system and reduction of direct intervention of the people, no significant deterioration of water quality was observed. After the commissioning of sewerage system, further improvement in water quality of the lake is expected.

Lessons Learned

Urban lakes are prone to reclamation and pollution. Sustainable management of lakes requires appropriate measures such as:

- An integrated approach for preparation of management plan.
- A common code training to project staff that covers diverse areas and interdisciplinary topics is necessary for proper formulation and implementation of the plan.
- Sustainable management of lakes/wetlands is directly linked to protection of its catchment. Application of "Prevention is better than cure" principal is a must for tackling the growing anthropogenic impact on the lakes/wetlands.
- Regulation of developments in the catchment to prevent destruction/modification of natural inlets to allow flow of rain water only from the catchment to the lake is required.
- In case of lakes of potable water source, rationalization of water withdrawal is necessary; withdrawal may be linked to highest level of water after the monsoons.
- A bio-physical barrier along the lakes may prevent encroachment and direct human intervention.
- A well coordinated and sustained environmental campaign is required to involve the people with the implementation of management plan.
- In India religious activities have a deep relationship with water resources. Immersion of idols in water is practiced throughout the length and breadth of the country, which is not complementary to the sustainable management of lakes. Idol immersion in potable water source could be prevented through public consent through dialogue and administrative support.
- Regular water quality monitoring of the indicative parameters is necessary.
- Lake/wetland conservation programs usually involve large financial investment but the return is not directly measurable, and thus difficult to convince the people regarding the investment need. The quantification of benefits, thus have to be linked to peoples health, and recreational benefits being derived by them due to conservation efforts.
- There is need to have comprehensive monitoring of lake water quality and

this long term data can provide us with a proxy understanding of the relative contribution of the rural and urban sections of the catchment.

- The urban pollution control measures have been taken care under the Bhoj Wetland Project while rural pollution measures need to be given due importance for reducing of silt and nutrient load.
- The agricultural runoff and siltation from the rural catchment will require a combination of technical and institutional innovations. These low cost and innovative mechanisms can be financed and monitored by the stakeholders benefiting from the lake.
- Agriculture is the predominant land-use in the catchment, with over 2/3rd land under farming. The problem of rural sector pollution may be minimized by adopting wetland friendly agricultural practices and technologies that are also user-friendly from the farmer's point of view. Care has to be taken to ensure that the options address the economic viability concerns of the farmers. Options for remediation include both measures at the source (the fields) and in transit (riparian zones and streams)

Conclusion and Recommendations

The Conservation and Integrated Lake Basin Management of Bhoj Wetland can be achieved through following measures to improve overall environmental scenario of catchment area and the upper lake:

- Constituted Lake Conservation Authority by State Government to sustain the conservation efforts made under the project.
- Declaration of buffer zone plantation area as well as the southern transitional zone of the Upper lake as a bird sanctuary, a first in state of Madhya Pradesh, to provide long-term protection to the plantations raised and to promote eco-tourism.
- The draft Regulatory Framework for Wetland Conservation has been prepared by Ministry of Environment and Forests, GOI in consultation with all experts covering all aspects of conservation and restoration of wetlands which is under consideration.
- Addressing these issues of agricultural runoff and siltation from the rural catchment will require a combination of technical and institutional innovations. These low cost and innovative mechanisms can be financed and monitored by the stakeholders benefiting from the lake. Bhopal Municipal Corporation is the primary stakeholder as the owner of the lake and it can represents the interests of the citizens of the city. In addition, major industries and other institutions can also contribute towards catchment protection. The process can be designed and facilitated by independent bodies like the LCA and other partners of their choice including Winrock International India.
- There is need for restriction on water withdrawal for potable as well as for other uses especially agricultural whenever there is less than average

rainfall. In this regard the post monsoon maximum water level of Upper lake at 507 m may be considered as the threshold limit.

- Stream bank erosion control at critical erosion prone areas along the river through loose boulder pitching.
- Removal of silt deposits from the problem areas to maintain proper gradient in the Kolans.
- Restriction of farming atleast upto 30 m on both sides of the river bank to minimize soil erosion from agricultural field.
- Promotion of community plantations in the catchment.
- Restriction of construction of water harvesting structures along the Kolans and its tributaries.
- Rationalization of water withdrawal for potable purposes throughout the year when the post monsoon maximum water level of Upper Lake become 507 m or less.
- Restriction on pumping of water from the Upper Lake for agriculture purposes.
- Gradually minimizing the use of Upper lake water for drinking purposes by exploring various alternatives.
- Creation/Identification of a singular focused nodal agency to coordinate with various department/local bodies [Bhopal Municipal Corporation, District Panchayat (Bhopal and Sehore), Agriculture Department, etc.] for maintaining status-quo in the catchment area of Upper Lake.
- There are 55 villages in the 361 sq km catchment area of Upper Lake of which very few villages are relatively large in terms of area and population. Majorities of them are engaged in agricultural and allied activities, there is very little diversification.
- About twelve villages are in the close proximity of the lake, while rest of the villages is either located along the Kolans or far away from the lake periphery.
- Soil erosion from these villages is the major problem. This problem is all the more serious because the forest cover in the catchment is thin. To tackle this problem of soil erosion tree plantation in the wastelands may be initiated.
- Crop mix and crop rotation is another safer method of improving the quality of land and thereby increasing the productivity of land.
- Detailed investigation is to be carried out to increase the catchment area by diverting the drainage from the other nearby catchment to maintain the constant inflow into the upper lake and to ensure availability of water for water supply to the city.
- The environmental and wetland friendly activities can be promoted and low density construction can be allowed in the rural catchment for conservation of Upper Lake. Proper planning is required to restrict the land-use and developmental activities in controlled manner.

- ✰ The provision of Bhopal Master Plan (2005) for no construction zone at a distance of 50 m from the fringe of upper lake should be strictly followed to prevent lake.
- ✰ Dairies located near upper lake should be rehabilitated elsewhere outside the catchment area.
- ✰ Buffer zone by planting appropriate species of plants in available open spaces should be created all around the Bhoj Wetland.
- ✰ Environmental impact assessment study should be conducted before implementing any major new recreational, tourism and developmental activities in and around Bhoj Wetland for taking necessary decision by technical committee as it is source of potable water to the city.
- ✰ Use of polythene can be prohibited in and around the lake as it is harmful for the aquatic animals.
- ✰ Lotus and Trapa cultivation should be completely banned/prohibited in the upper lake as cultivators are using pesticides and insecticides for above cultivation which deteriorates the water quality of the lake. Besides this, alternative job for the cultivators should also be explored to protect their livelihood.
- ✰ Maintenance of gabion/CAT structures constructed under Bhoj Wetland can be handed over to local village community/village/district administration and regular cleaning of silt deposited on the upstream of structure should be done through villagers for effective functioning.
- ✰ Regular weed removal from Bhoj Wetland should be carried out through local body to improve the water quality.
- ✰ Blue print of Bhoj Wetland can be prepared to disseminate the information and create awareness among the people for the conservation and management of lake.
- ✰ In sum, the topographic situation of the lake suggests that water within the lake moves predominantly from west to east and not in the reverse direction. Therefore the urban pollution load can't be shifted to rural fed lake area. The existing pollution in the lake is substantially contributed by the rural catchment. While nitrate levels are not a problem while they lead to increased microbial activity. Higher Coliform count and turbidity lead to higher flocculation and disinfestations costs and clog sand filters with algae. Therefore to reduce the cost of the treatment for potable water, it is essential to control the productivity of the lake by controlling and checking of inflowing organic and inorganic matter through point and non-point sources. The decrease in the productivity shall result in the decrease in the turbidity of water and increase in the transparency of water column, thereby reducing the treatment cost of water. Accordingly, nutrient and organic matter runoff from the rural catchment needs to be addressed for maintaining water quality and sustaining the lake into the future, particularly as urban point sources are tackled.

Acknowledgement

Madhya Pradesh District Gazetteers of Sehore and Bhopal - P.N.Shrivastav and S,D Guru-1989.

Ex-Post project evaluation study conducted by Japan Bank for International Cooperation, Japan (2006).

Bhoj Wetland Project, EPCO, Bhopal.

Economic Valuation of Bhoj Wetland conducted by Indian Institute of Forest Management, Bhopal.

Master Plan of Bhopal (2005) prepared by Town and Country Planning Department, Govt. of M.P.

17

Fuel Oil from Wastelands: Jatropha Project

Kumud Dubey, Md Rizwan Khan, Anubha Srivastava and V.K. Singh

INTRODUCTION

Jatropha curcas is a multipurpose plant with many attributes and considerable potential. It is a tropical bush or small tree that can be grown in low to high rainfall areas and can be used to reclaim land, as a hedge and/or as a commercial crop. Thus, growing it could provide employment, improve the environment and enhance the quality of rural life. It produces 35-40 per cent seed oil on weight basis. This oil is a renewable resource, and a safe source of energy. It is a viable alternative to diesel, kerosene, LPG, furnace oil, coal and fuel wood. In recent times, an excellent awareness has been created both by govt. and non-govt. organizations to popularize *Jatropha curcas* as a bio -fuel crop. Since the crop is hardy and has wide adaptability to varied climate and soil, it is proposed to promote Jatropha plantation in wastelands of the country (Annons,2001, Singh et. al, 2005., Openshaw, 2000). In addition to its oil, being a potential substitute to diesel, several other properties such as wide environmental tolerance, grown on any type of soil, well adopted to various kinds of wastelands, easy propagation through seeds/cuttings, requires minimal after-plantation care, lesser gestation periods, not grazed by animals even during the times of drought, etc., strengthens its case for promotion in wastelands. A mission mode approach is proposed to be adopted for Jatropha plantation in potential states/districts of the country with a view to:

- ☆ Creating rural employment generation,
- ☆ Greening wastelands,
- ☆ Drought proofing,
- ☆ Reducing growing dependence on imported petroleum products,
- ☆ Achieving Bharat-III norms by 2007 and Bharat-IV norms by 2010 in respect of vehicular emissions corresponding to Euro-III and Euro-IV norms of emission,
- ☆ Securing energy security for the country in general and for rural areas in particular, and promoting organic farming.

India ranks sixth in the world in terms of energy demand accounting for 3.5 per cent of world commercial energy demand in 2001. The energy demand is expected to grow at 4.8 per cent. The demand of Diesel (HSD) is projected to grow from 39.81 million metric tons in 2001-02 to 52.32 million metric tons in 2006-07 @5.6 per cent per annum. Our crude oil production as per the Tenth Plan Working Group is estimated to hover around 33-34 million metric tons per annum Our dependence on import of oil will increase in the foreseeable future. The Working Group has estimated import of crude oil to go up from 85 million metric tons per annum to 147 million metric tons per annum by the end of 2006-07 correspondingly increasing the import bill from $ 13.3 billion to $ 15.7 billion at today's prices. Transport sector remains the most problem sector as no alternative to petroleum-based fuel has been successful so far. Hence petroleum based fuels especially petroleum diesel (HSD) will continue to dominate the transport sector in the foreseeable future but its consumption can be minimized by implementation of Biodiesel. Since edible oil demand is higher than its domestic production, there is no possibility of diverting this oil for production of biodiesel. Fortunately there is a large junk of degraded forest land and unutilized public land, field boundaries and fallow lands of farmers where non-edible oil-seeds can be grown. There are many tree species which bear seeds rich in oil. Of these some promising tree species have been evaluated and it has been found that there are a number of them such as Jatropha curcas (Ratanjyot) and Pongamia Pinnata ('Honge' or 'Karanja') which would be very suitable in our conditions (Katwal and Soni, 2003).. However, ***Jatropha curcas* (*Ratanjyot*)** has been found most suitable for the purpose for the reasons summarized below:

- ☆ It can be grown as a quick yielding plant even in adverse land situations *viz.* degraded and barren lands under forest and non-forest use, dry and drought prone areas, marginal lands and as agroforestry crop. It can be planted on fallow lands and along farmers field boundaries as hedge because it does not grow too tall as well as on vacant lands alongside railways, highways, irrigation canals and unused lands in townships etc. under Public/Private Sector Undertakings.
- ☆ The seeds of Jatropha are available during the non-rainy season, which facilitates better collection and processing. The cost of plantation is largely incurred in the first year and improved planting material can make a huge difference in yield.

- ☆ Raising Jatropha plant and its maintenance creates jobs for the rural poor, particularly the landless, in plantation and primary processing through expellers.
- ☆ It has multiple uses and after the extraction of oil from the seeds, the oil cake left behind can be used for Bio Gas production and is an excellent organic manure, the bio mass of Jatropha curcas enriches the soil and it can also be put to other uses.
- ☆ Retains soil moisture and improve land capability and environment.

Jatropha adds to the capital stock of the farmers and the community, for sustainable generation of income and employment.. It will use lands which are largely unproductive for the time being and are located in poverty stricken areas and in degraded forests. Planning commission has asked states to grow more of Jatropha on wasteland and semi rain fed areas.

Researches are going on various aspects of the Biodiesel plant *i.e. Jatropha curcas.* Our centre is working on different aspects of the selection of superior planting material for quality and quantity of oil and their suitability for agro climatic zone of eastern U.P. The present study is on the relative growth performance of Jatropha (*Jatropha curcas*) provenances at nursery stage. The seeds of these provenances were collected from different parts of India.

Materials and Methods

The experiment was conducted at Padilla Research Nursery of the centre. Total twenty-three provenances were collected from different part of India. The seeds of these provenances were procured from the different research organizations from all over India working under the National Networking Programme on Jatropha. The seeds were cleaned manually. Before sowing, the seeds were soaked in water for 12 hours at ambient temperature and sown in polythene bags which were already filled with soil, FYM and sand in 1:1:1 ratio in the month of February. The experiment was performed in three replication and in each replication there were eighteen plants. The germination were started after fifth day and completed after fifteen days. The plants were irrigated as and when they required. The weeding and hoeing were done regularly. The data for germination percent were recorded. The growth parameters *viz.* plant height and collar diameter was recorded after three months of sowing. The survival percentage was also recorded after three months of sowing.

Results and Discussion

The growth data *viz.* plant height and collar diameter were sownin Table 17.1. It is revealed from the Table 17.1 that highest germination (79.6 per cent) and Survival percent (78.8) was found in PJ Sel-2 variety of Panthnagar University followed by PJ Sel-1 variety of Panthnagar (germination 77.2 per cent and survival percentage 75.8 per cent), Jhansi (Germination-67.6 per cent and survival 66 per cent) and being lowest with TNMC-5 variety of Coimbatore (Germination- 23.4 per cent and survival-22.2 per cent). Where as the highest plant height was obtained from the planting material procured from PKVJ-MKU-1 variety of Akola (44.36 cm) followed by

Table 17.1: Relative growth performance of different provenance's

Sl.No.	*Provenance*	*Germination Per cent*	*Survival Per cent*	*Height after 90 DAS (cm)*	*Collar Dia. after 90 DAS (cm)*
1.	Kanpur	67.1	63.8	32.63	1.59
2.	Jhansi (NRCJ-41)	62.5	58.5	38.46	1.52
3.	Jhansi (NRCJ-42)	65.3	61.33	37.56	2.06
4.	Jhansi (NRCJ-47)	36.15	32.3	37.66	1.39
5.	Jhansi (NRCJ-53)	67.6	66	32.26	1.61
6.	Akola PKVJ-MKU-1	76	75	44.36	1.72
7.	Akola PKVJ-DHW-1	60.8	60	25.1	1.13
8.	Rahuri (RJ-117)	48.5	46.5	33.53	1.33
9.	Panthnagar (PJ Sel-1)	77.2	75.8	39.96	1.64
10.	Panthnagar (PJ Sel-2)	79.6	78.8	44.1	1.75
11.	Jabalpur (TFRI-I)	53.75	50.41	37.06	1.83
12.	Jabalpur (TFRI-II)	63	59	39.23	2.19
13.	SFRI, Jabalpur (Indore)	24	22.8	28.1	1.08
14.	SFRI, Jabalpur (Sagar)	67.4	65.8	29.63	1.12
15.	Jabalpur (Silviculture nursery)	55.4	53.8	39.53	1.32
16.	Jabalpur (Barah Experimental area)	58	56.6	38.13	1.22
17.	Allahabad (Shankargarh)	57.4	55.4	35.23	2.15
18.	Coimbatore (TNMC-2)	31	31	32.1	1.37
19.	Coimbatore (TNMC-3)	44.4	43	34.3	1.67
20.	Coimbatore (TNMC-4)	36.6	34.8	27.93	1.55
21.	Coimbatore (TNMC-5)	23.4	22.2	26.46	1.35
22.	Coimbatore (TNMC-7)	63.6	62.4	38.46	1.66
23.	Coimbatore (TNMC-22)	65	64.2	38	1.57

Panthnagar (44.1 cm) and was lowest with PKVJ-DHW-1variety of Akola (25.1 cm). Whereas the highest collar diameter was obtained from TFRI-II variety of Jabalpur (2.19 cm) followed by Allahabad (Shankargarh) (2.15cm) and was lowest with State Forest Research Institute, Jabalpur (1.12 cm). Since the concept of substituting biodiesel has gained widespread attention in India. Moreover bio-Diesel is biodegradable, Non-Toxic, free from sulfur and aromatics and ecofriendly. It is a renewable resource based on oil crop that are grown anew each year. It can be produced domestically, reducing our country's dependence on imported oil, thus huge savings on imports. Since Jatropha curcas is well-established plant in India, it can be planted as bio diesel crop. It produces oil-rich seeds, is known to thrive an eroded lands, and to require only limited amount of water, nutrient and capital inputs. This plan offers the option both to cultivate wastelands and produce vegetable oil suitable for conversion to biodiesel. Recently the Indian Government as well as some State Governments has expressed their support for bringing marginal land and wastelands, which cannot be used for food production, under cultivation for this purpose. It helps to increase rural incomes, self-sustainability and alleviate poverty and hence this plant offers the option both to cultivate wastelands and to produce vegetable oil suitable for conversion to bio-diesel and other economically important by products as an alternative resources of income (Francis, G. *et al.*, 2005). Thus through Jatropha plantations, three major objectives, *viz.* wasteland reclamation, fuel production, and socio-economic development of degraded areas can be achieved simultaneously.

References

Anon., 2001. *The Wealth of India*, CSIR, New Delhi, **5**: 293-295.

Chaudhari, D.C. and Joshi, D.N., 1999. *Jatropha curcas*: a multipurpose species for economic prosperity and wasteland development. *Advance in Plant Science Research in India*, **9**: 35-39.

Francis, G., Edinger, R. and Becker, K., 2005. A concept for simultaneous wasteland reclamation, fuel production, and socio-economic development in degraded areas in India: Need, potential and perspectives of Jatropha plantations. *Natural Resources Forum*, **29(1)**: 12.

Katwal, R.P.S. and Soni, P.L., 2003. Biofuels: An opportunity for socio-economic development and cleaner environment. *Indian Forester*, p. 939-949.

Openshaw, K., 2000. Review of *Jatropha curcas*: An oil plant of unfulfilled promise. *Biomass and Bioenergy*, **19**: 1, 1-15.

Singh, V.K., Dubey, K. and Srivastava, A., 2005. *Jatropha: Ek Bahupayogi Paudha*. CSFER Allahabad Publication.

18

Solid Waste Treatment and Disposal Options

Sutapa Bose, Gurmeet Singh & Anil K. Gupta

INTROUCTION

Over the past several decades, the rapid globalization of the world economy has led to phenomenon levels of economic growth, particularly in the northern hemisphere. However, this economic growth has brought with its side effect, including the generation of an increasing volume of hazardous waste. India is one of the few developing countries, trying to raise its position in global market. Not so long ago, man was convinced that science and technology would create a paradise on earth, which it succeeded to certain extent. But, unfortunately there has been another side of the coin. Industry is often considered as environment's major enemy. Industry is not only criticized for processes that pollute, but also for creating products that are wasteful of scarce resources. The age of industrialization has brought severe damage to the environment by degrading land, water and air. This environmental degradation is directly related to population growth, industrialization, rise of living standard, urbanization etc. Nearly every anthropogenic activity leaves some kind of wastes. Households create ordinary garbage; industrial and manufacturing processes create solid and hazardous waste.

Waste can be defined as any movable material, that is perceived to be of no further use and that is permanently discarded. Under London Convention 1972, Waste has been defined broadly to include the 'material and substances of any kind, form and description.' Basel Convention, 1989, on the other hand defines waste by reference to there end use; "they are the substances or objects, which are disposed of, or are intended to be disposed of, or are required to be disposed of by the provision of national law." Once in the environment, wastes frequently caused damage to ecosystems and human health and therefore act as pollutants.

Around most town and cities in India, improper dumping of waste as plastic bags, domestic wastes etc alongside road is quite common. Burning of waste in order to reduce the volume is also a prevalent practice, which contaminates the atmosphere as well. The lack of space for dumping the solid waste has become a serious problem in several cities and towns all over the world. Modern methods of disposal such as, incineration and the development of sanitary landfills, etc are now attempting to solve these problems. The disposal of solid wastes has to be considered as a part of an integrated waste management plan with an 'integrated life cycle approach'. The methods of collection, processing, resource recovery and the final disposal should be synchronized to achieve a common objective, i.e. sound disposal and management of the waste.

Source of Solid Waste

Waste can be defined according to their source of origin. They can be mining, agriculture, industrial, municipal wastes and sewage sludge, among them industrial process wastes encompass a very wide range of materials and may include general factory rubbish, packaging materials, organic wastes, acids, alkalis and metalliferous sludge It creates problems because it is unwanted.

The waste can be classified according to their sources,

Domestic municipal source: The domestic municipal waste is mainly of organic in nature.

Non domestic source: This fraction comprises the industrial pollution include thermal power plants, brick kilns, hot mix plants, industrial units, agricultural sectors and mining sectors.

Industrial waste is more troublesome than other wastes. It consists of toxic inorganic, organic and high concentration of heavy metals, which are causing harmful health effect to the living organisms and ecosystem (Subrahmanyam, 1991). Solid wastes may be hazardous or non-hazardous. 'Hazardous wastes' could be defined as "The wastes other than radio active wastes which by reasons of their chemical reactivity or toxicity, explosive, corrosive or other characteristics causing danger or like to cause danger to health or the environment, whether alone or coming into contact with other wastes, are legally defined as hazardous in the state in which they are generated or in which they are disposed of or through which they are transported" [UNEP, 1989].

Domestic Municipal Wastes

Those wastes generated by households and derived from different commercial units, are come under domestic municipal wastes. It is mostly non-hazardous solid waste of city. The quality of municipal wastes depends on the growth rate of industrialization as well as socioeconomic conditions. Per capita waste generation varies between 2.75-4.0 kg per day in high-income countries, but is as low as 0.5 kg per day in these countries with lowest income (UNEP, Environmental Data Report, 3rd edn., Blackwell, Oxford,1992). Domestic municipal solid waste contained a wide variety of materials. It can be contained food waste, which classified as wet garbage as well as paper, plastic, glass, cardboard, aluminum foil etc., which is classified as dry garbage.

Non-domestic Wastes

It covers various fields, agriculture, industrial, mining, sludge etc., which are both hazardous and non-hazardous. Waste produced by agricultural activities comprises, animal slurries, end of spray residues and tank washing followed by pesticides use and empty plastic packing.

Sewage sludge is produced as a result of treatment of industrial and domestic wastes. Sewage sludge may be contained heavy metals, water-soluble organic-rich pesticide, oil and greases, which make the sludge hazardous. Industrial waste comprises a wide range of material, organic wastes, acid, alkalis, metals etc. Mining wastes arise as by-product of the extraction processes, in which a significant proportion is regarded as hazardous and, such wastes require a special treatment and disposal.

Hazardous Wastes

A waste is classified as hazardous if it exhibits any four of its primary characteristics based on the physical and chemical properties of toxicity, reactivity, ignitibility and corrosively. In addition to this waste products that are either infectious or radioactive are also classified as hazardous.

Table 18.1: The composition of typical urban refuse derived from an industrial country

Component	*Proportion % by Weight*
Paper	35
Garden waste	16
Food wastes	15
Metals	10
Glass	10
Plastic, rubbers and leathers	7
Rags	2
Miscellaneous	5

(UNEP, Environmental Data Reort, 3rd edn., Blackwell, Oxford, 1991)

Sewage is considered as low hazardous waste, which consists of outflow from domestic and industrial premises and in some cases the run-off from the road. Only a very small fraction of sewage (0.05 per cent) is waste materials, the rest being water. Despite its apparently low waste content, the discharge of untreated sewage into surface water can lead to gross pollution.

Assessment of the amount of high hazardous waste generated on a global basis is problematic. This is in part because of inadequate record keeping, but also because there is no uniformly accepted definition of high hazardous waste. There are strong indications that the production of high hazardous wastes is vast and expanding, both absolutely and as a portion of industrial waste as a whole. Global production of hazardous wastes is estimated to be at least 3.38X 10^{11} kg a^{-1}, about 80 per cent of which is generated in the USA (*UNEP,* Environmental Data Report, *3rd edn., Blackwell, Oxford, 1991*). The remedial measures of hazardous wastes are likely to be extremely costly.

Solid Waste Management: Legal Approach

Prior to 1970 the industrialized north disposed of these waste within their own countries, often with little regard to the environmental impact of disposal. However as incident of improperly disposed of waste, such as *"The Love Canal Affair"*, in the United States began to create serious health and ecological problems, public awareness of the consequences of improper disposal increased. Eventually this awareness led people in industrialized countries to expressing an increasing unwillingness to have waste disposed of at home [The NIMBY (not in my backyard) syndrome].

International regulation regarding the solid waste management began in the early seventies with the adaptation of two treaties, which prohibited the disposal at the sea of certain type of waste (The Convention on the Prevention of Marine pollution by Dumping of waste and other matter, London 1972; and International convention for the prevention of pollution from ships [MARPOL] 1978.)

Table 18.2: Share of continents (developing countries) receiving hazardous waste from developed countries (Santra, 2001)

Hazardous Waste Shipment Status (1987-1988)	
Continent	*Amount (tons)*
Africa	507668.7
North America	54386
South America	29,758
Asia	6209
Europe	3052818

At global level, no UN or other body has overall responsibility overall responsibility for waste which has led a fragmented, ad hoc and piece meal international response. The Stockholm Conference did not really grapple with the issue of waste, as in 1972 the problem was not as severe as it is now. Without specifically mentioning waste, Principle 6 of the 1972, declaration called for the discharge of toxic and other substances to be halted. The 1982, world Charter for Nature called for the 'special precaution' to be taken to prevent the discharge of radioactive or toxic waste, but it kept mum regarding the minimization of generation of such type of waste. At UNCED, the issue of waste was addressed in some details in Agenda 21 with development of proposal, including target and timetables, for the management of hazardous and other waste, and radioactive waste. Agenda21 distinguishes between Hazardous waste, solid waste (including Sewage), and radioactive waste. Principle 14 of Rio declaration limited itself to calling for the effective co-operation to discourage, or prevent the relocation of transfer of to other states of any activity and substances that caused severe environmental degradation or are found to be harmful to human health. The goal of Basel convention was ESM [Environment Sound Management], which addresses " Integrated Life Cycle approach " which involves strong controls from the generation of a hazardous waste to its storage, transport, reuse, recycling, recovery and final disposal. In other words, the minimization of hazardous waste should be done whenever it is possible. Despite of these convention and global consensus, many of the developed countries turned a deaf ear towards the problem of solid waste disposal

ISWA (International Solid waste Association) was established in Essen, Germany in 1970 by founding members from Germany, Austria, Switzerland, France, Italy and UK. It is a non-profitable, non-government organization, having headquarters in

Copan Hagen, Denmark. It has membership strength of 75 countries and it acts a component partner of UNEP (http://www.test.iswa.org/pdf/AnnualREPORT2002.pdf). It has the objectives to disseminate information on the all aspects of solid waste management as well as to exchange information in this regard.

Waste Treatment and Disposal

The three R's Reduce, Recycle and Recover are important management, are the important pillar of the solid waste management. The most effective strategy for reduction of solid waste generation is the reduction at generation. Options for reuse include purification followed by recycling. This approach is frequently applied to solvents, as recovery of pure material from waste solvent is often achievable by distillation. An alternative approach is to use the waste from one process as a feedstock for another. There are instances, where the waste from one process can be used to treat the waste from other. Reuse within the facility that generated the waste is desirable as the need for transport is minimized. Alternatively, the wastes generated by one manufacturing company may be used directly by another. In some cases establishment of "waste exchanges" has encouraged this. The waste generated from thermal power plant (fly ash) has been successful used for the brick making as well as cement manufacturing.

Sewage treatment is primarily aimed at lowering the pathogen content of the wastes. Additional objectives include a decrease in its biochemical oxygen demand (BOD) and solid contents. These objectives may be achieved in a series of stages (Figure 18.1).

Primary treatments remove the large objects within the raw sewage. The processes used are mechanical. Grit separation is achieved using gravity under condition where the less dense organic matter remained in suspension. In all but most rudimentary plants, the effluent from preliminary treatment is often subjected to primary treatment. During the process the sewage is allowed to slowly traverse a tank, allowing about half of

Secondary treatment is a biological process. Three designs of reactor are commonly used, trickling filer, activated sludge tank and oxidation pond. The last of these only appropriate in warm climate.

The preliminary, primary and secondary treatments outlined above are highly successful. When in a combination they are capable of producing in an effluent with low BOD, low suspended solid and low total solid. Pathogen populations are also greatly reduced. The final effluent from the treatment works is discharged in a river. Tertiary treatment is seen as desired in location where the degree of dilution of the effluent is small or where potable water is to be drawn down stream for treatment and distribution. The purpose of this treatment is to further reduce BOD and concentrations of suspended solid, nutrients, toxicants (heavy metals, phosphorus organics). Wide ranges of technology have been developed to facilitate the desired improvement. After tertiary treatment, the sludge can be disposed in the sea, incinerate or dumped on the land (landfill), or can be used on land as a fertilizer or conditioner. The last of these is restricted by both transport cost and heavy metal burden of the sludge. Landfill

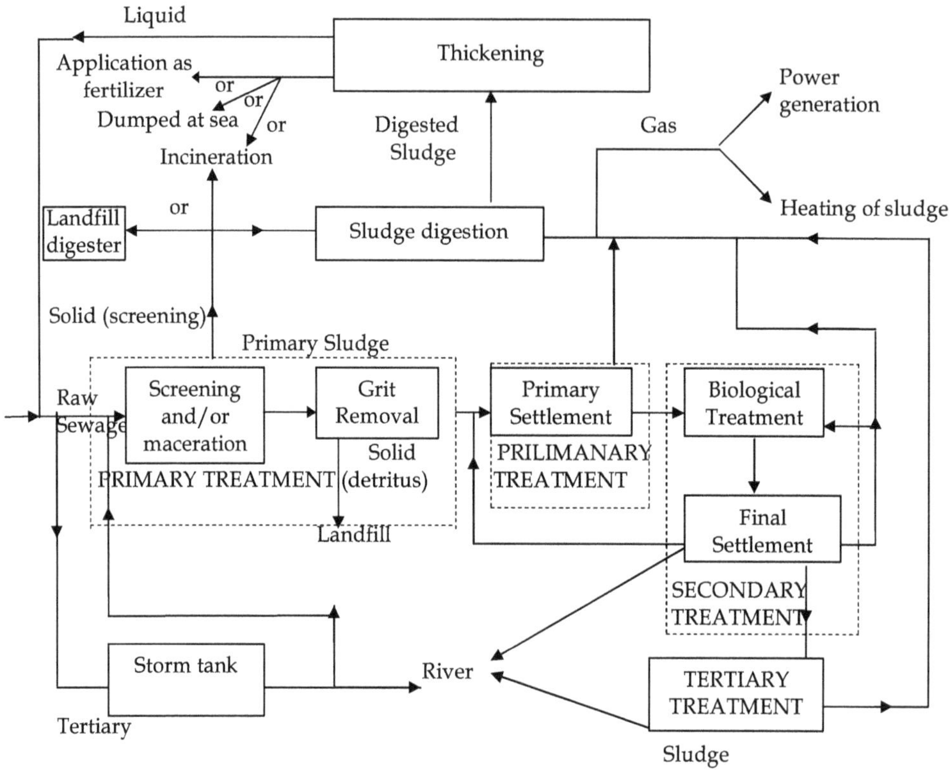

Figure 18.1: The sewage treatment process.

sites are increasingly scares and dumping at the sea is becoming restricted by legislation; in the USA this practice was banned in 1991, and in the UK it was ceased in 1998. Therefore he percentage of sludge that is sent to incinerate appears likely to increase, at least in the most developed countries.

The sludge is digested and the settle sewage then enters secondary treatment.

Strategies for the treatment of highly hazardous wastes can be divided into those aimed at reuse, at destruction or immobilization. Destruction of high-hazard waste is only applicable to those that are hazardous by virtue of molecule that they contained rather than their constituent elements. For example cyanide ion (CN^-) is found in wastes from metal processing industries. It is highly toxic even though both of its constituent elements are essential for life. It is toxic by virtue of its affinity for the active sites of enzymes involve in respiration. Waste containing this species may be detoxified by treatment with chlorine (Cl_2). Unlike cyanide metal cannot be destroyed. The contaminants vary from source to source; however they may include Zn, Cu, Ni, Cd, Cr, Pb and Hg. These and many other heavy metals can be largely removed from the water by the addition of anions that cause the precipitation of metals as an insoluble salts. But these processes are highly cost effective.

Incineration of low and high-hazard wastes shares the same draw back, namely potential air pollution. The solid waste can be disposed of by mean of various thermo-chemical processes such as incineration, pyrolysis and wet air oxidation. Incineration finds greater advantage in the disposal of materials, such as hospital wastes, that may be contaminated with pathogens.

Biological methods of hazardous waste disposal have been used for sometime(*Bose et al 2005, Chandrayan et al., 2005, Singh and Bhattacharyya 2005*). Bioremediation including phytoremediation is one of the low cost technologies for the removal of hazardous wastes (*Bose and Bhattacharyya 2008, Bose et al., 2008 a, b, c*). In one system, known as land farming, oily wastes are spread onto the soil. Decomposition may be enhanced by the addition of organic fertilizers and the periodic disturbance of the land using conventional agricultural implements. This generates high conditions for the breakdown of the wastes by the naturally occurring soil microorganisms. How ever interesting to note that disposal of sewage sludge on farms has been banned in Switzerland in 2005. The move is in response both to a reduction in demand for sludge and increasing concern about the residues in the sludge, particularly synthetic hormones and some pharmaceutical compounds.

Wastes that are neither recycled nor destroyed must be disposed of. This can be done with much greater safety if the waste is immobilized first. The technologies used to do this involve either the incorporation of waste into a solid matrix or its encapsulation within an impermeable polymeric cover.

In addition to immobilization these processes are variously referred to as stabilization, solidification or fixation.

Historically the bulk of hazardous wastes have been disposed of to the land, often with little or no pretreatments. For example, in 1985 the UK disposed of about 2.75×10^9 kga^{-1} of chemical wastes was sent to landfill; this compares with a total about 4.2×10^8 kga^{-1} was treated chemically, fixed or incinerate. Modern secure landfill sites are located where the probability groundwater contamination is likely zero. They are covered and lined with impermeable membranes and leachates are collected, monitored and treated (Figure 18.2a-b).

Other procedures that are used for the disposal of hazardous wastes include dumping at sea. For example, in 1985 the UK disposed of about 2.3×10^8 kga^{-1} of chemical wastes in this way. However there have been political moves to curb this practice. It was agreed by the 13th Consultative Meeting of the London Dumping Convention (1972) that all sea dumping of non-inert industrial waste should cease by 31 December 1995. However, despite of deadline, ocean and sea are considered as a good alternative, though not eco-friendly for dumping of waste. Hazardous wastes have also been disposed of by placement at depth within the earth, well out of reach of potable aquifers. This has been done within disused mines and by deep-well injection. These are the methods of long term storage or ultimate disposal and have been widely used in US for steel mill waste, paper mill waste, chemical industrial waste, nylon waste and complex wastewater (Tofflemire and Brezner, 1971).

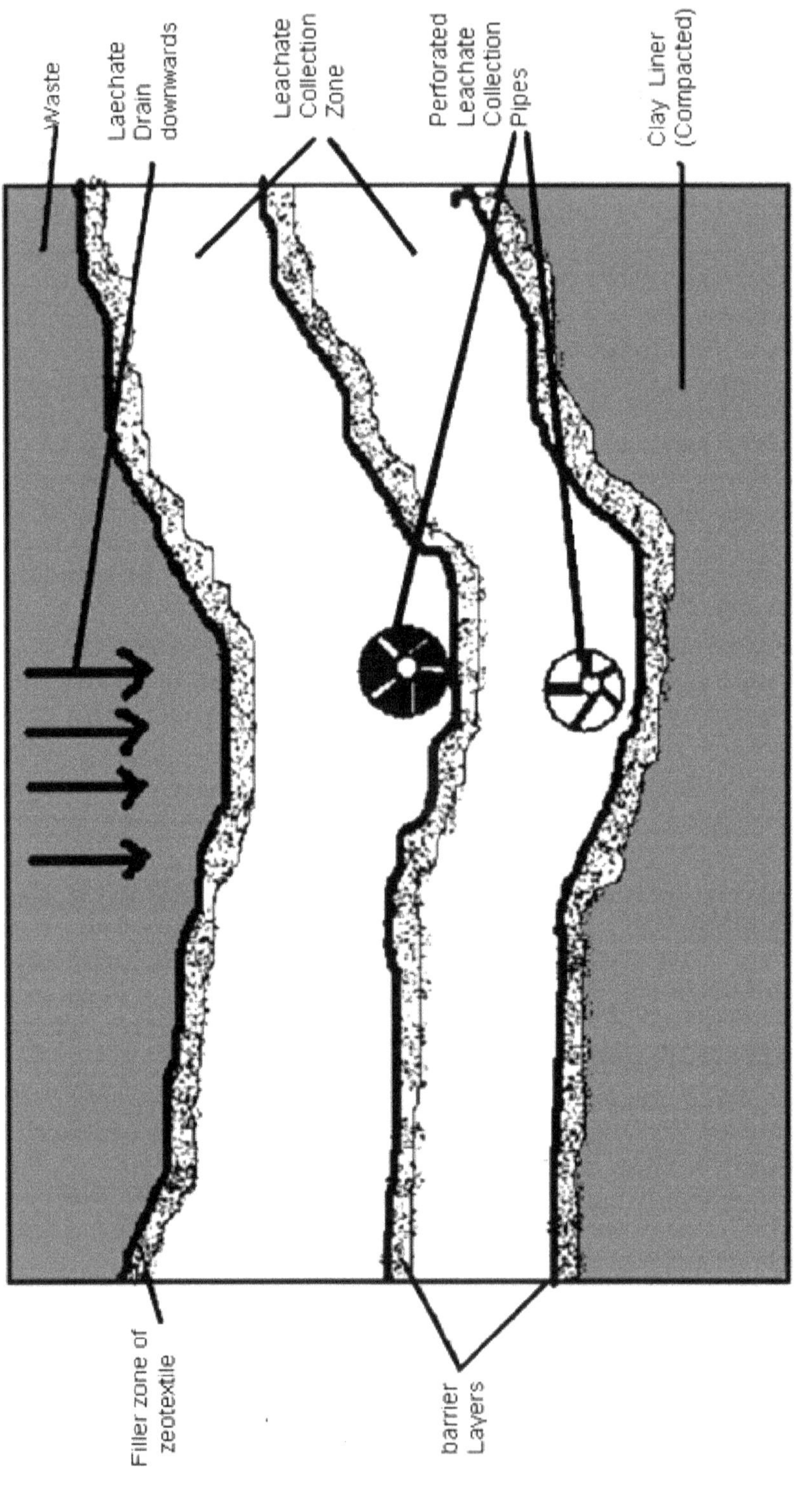

Figure 18.2(a): System suitable for the lining.

Figure 18.2(b): System suitable the capping of secure landfill sites.

Private Sector Participation in Solid Waste Management

Participation of private companies in solid waste management has been successful in many countries. The reason behind success is the failures of municipal systems to provide adequate services, or due to the pressure from national governments and international agencies. An important factor in the success of private sector participation is the ability of the client or grantor - usually a municipal administration - to write and enforce an effective contract. Awareness of local governing bodies is also an important factor for judging the bids from the private sector are reasonable as well as their role is justifiable. Three key components of successful private sector participation are **competition, transparency and accountability**. As an alternative to large (often international) companies that can provide most or all of the solid waste services in a city, microenterprises or small enterprises (MSEs) can be involved. They often use simple equipment and labour-intensive methods, and therefore can collect waste in places where the conventional trucks of large companies cannot enter. These MSEs may be started as a business, to create income and employment, or community members who wish to improve the immediate environment of their homes may initiate them. In Indian towns, Solid Waste Management, usually, has been a function of the local civic authorities. The issue has gained prominence in recent times, as there is a growing realization everywhere about the need and utility of environmental sanitation. Endorsing the recommendations made by the committee on SWM constituted by it, Honorable Supreme Court of India, has issued strict instructions to the urban bodies for compliance within a stipulated time frame The Sewage treatment plant in Delhi, managed by Suez-Decrement of France has been a successes throughout.

Public Awareness about Solid Waste Management

In any country, developed or developing, there are social or cultural norms accepted only by the society. Such norms affect designs of solid waste management systems Where the society allows only a certain social class or group to deal with solid waste, the availability of work force for solid waste collection and disposal becomes constrained by this rule. In some countries, directly handling human waste is a traditional taboo, which then prohibits the application of co-composting of refuse and human waste. The lack of understanding of local cultures and ways of life by the external support agency is often a cause of failure of a collaborative project.

Effective management of solid waste requires the cooperation of the general public. Lifting the priority of, and allocating more resources to, the solid waste management sector needs the support from decision makers. It is, therefore, important to ensure that public and decision makers' awareness activities are incorporated into the external support package. The aim of these activities is normally long term and it takes some momentum to build up before the effects are realized. But, once the interests of the public and decision makers in improving solid waste management are created, the sustainability of solid waste management projects will be significantly improved

At dumpsites, transfer stations, and street refuse bins; waste picking or scavenging activities are common scenes in developing countries. People involved have not received school education and vocational training to obtain knowledge and

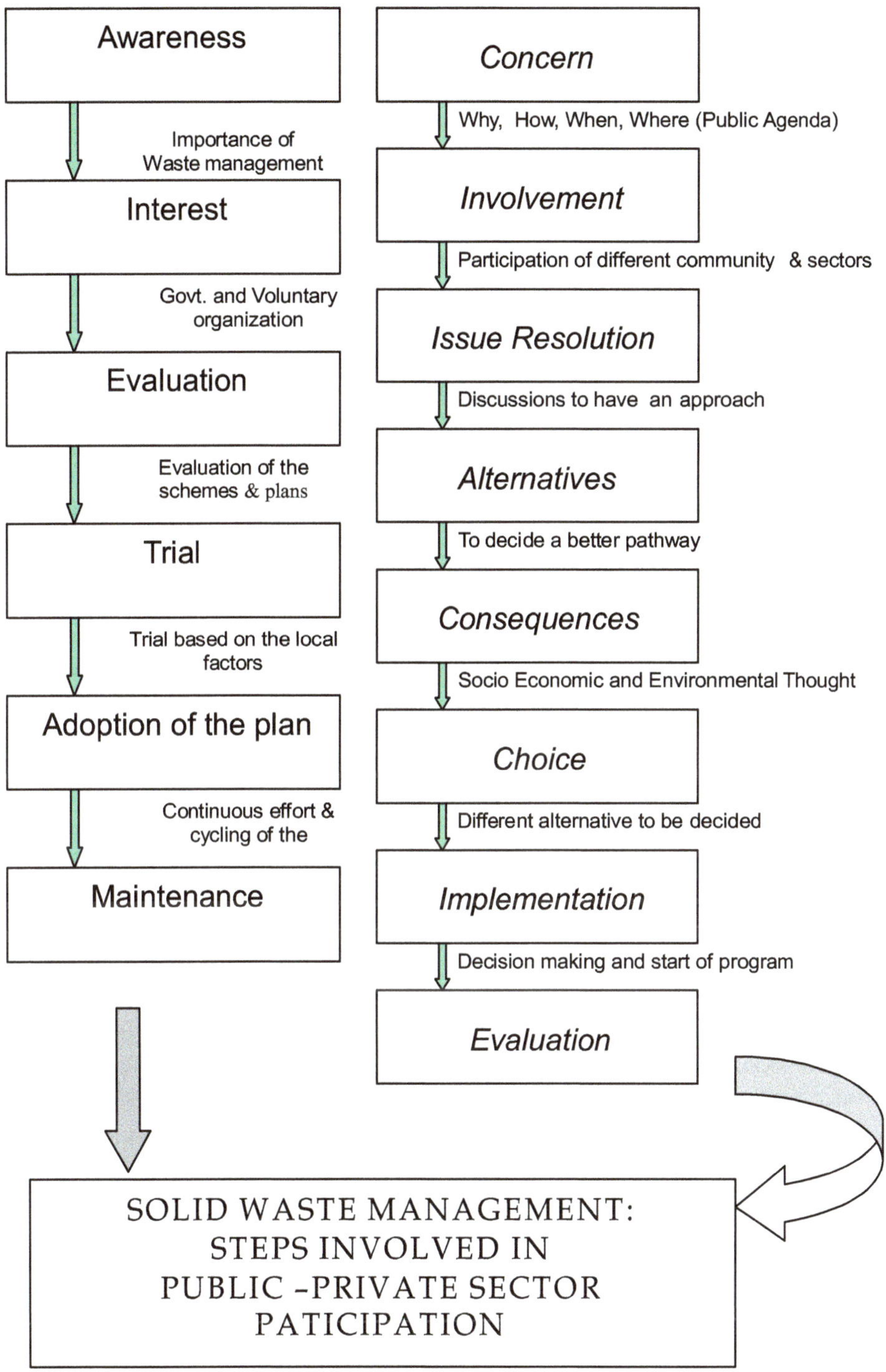

Figure 18.3: Public-private participation in solid waste management.

skills required for other jobs. They are also affected by limited employment opportunity available in the formal sector. The existence of waste pickers/scavengers creates often an obstacle to the operation of solid waste collection and disposal services. However, if organized properly, their activities can be effectively incorporated into a waste recycling system. Such an opportunistic approach is required for sustainable development of solid waste management programmes in developing countries.

Conclusion

Waste is said to be "a mirror of the society" since Waste generation and disposal reflect a range of aspects of the society, such as its economic, historical, cultural, and environmental components. Solid waste problems in developing countries include underdeveloped Waste collection systems and inappropriate final disposal, which is causing environmental pollution and placing an additional burden on those already suffering from urban poverty. Moreover, recent levels of economic growth and the subsequent introduction of mass consumption lifestyles from the developed countries are aggravating these problems. Capacity development support in SWM constitutes a key concept in sustainable management of the ecosystem. Appropriate treatment and disposal of waste can greatly ameliorate their environmental impact. However this 'end of pipe' technology cannot reduce the amount of waste generated. This can be achieved by in process modifications targeted as cleaner production. The concepts of cleaner production involve the application of integrated strategies aimed at avoiding unnecessary waste production and ensuring that the remaining waste produced is innocuous. This concept must be applied through out the life cycle of a product, from the extraction of the raw materials from which it is made through to its ultimate disposal.

References

Bose, S. and Bhattacharyya, A.K., 2008. "Heavy metal accumulation in wheat plant grown in soil amended with industrial sludge". *Chemosphere*, **70(7)**: 1264–1272.

Bose, S., Vedamati, J., Rai, V. and Ramanathan, A.L., 2008. "Metal uptake and transport by Tyaha angustata L. grown on metal contaminated waste amended soil: An implication of phytoremediation". *Geoderma*, **145(1-2)15**: 136-142.

Bose, S., Chandrayan, S., Rai, V., Bhattacharyy, A.K. and Ramanathan, A.L. 2008. "Translocation of metals in pea plants grown on various amendment of electroplating industrial sludge". *Bioresource Technology*, **99(10)**: 4467-4475.

Bose, S., Jain, A., Rai, V. and Ramanathan, A.L., 2008. "Chemical fractionation and translocation of heavy metals in *Canna indica* L. grown on industrial waste amended soil". *Journal of Hazardous Materials*, **160(1)**: 187-193.

Bose, S. and Bhattacharyya, A.K., 2005. "Land disposal of industrial solid wastes of Wazirpur, Delhi- Aspects of available calcium and magnesium". *International Journal of Pollution Research*, **24(4)**: 853-858

Chandrayan, S., Bose, S. and Bhattacharyya, A.K., 2005. "Effect of urban industrial solid waste application on soil available phosphorous and wheat plant response", Proceedings of International conference on *'Soil, Water and Environmental Quality-issues and strategies'*, Indian Society of Soil Science, New Delhi, Jan 28th to 1st Feb. 2005.

Santra, S.C., 2001. *Environmental Science*, New Central Book Agency (P) Ltd. Calcutta, pp. 1244.

Singh, G. and Bhattacharyya, A.K., 2005. "Land disposal of Urban Industrial Solid waste of Wazirpur, Delhi –Status of plant available sulfur". *Pollution Research,* **24(4)**: 309-314.

Tofflemire, T.J. and Brezner, G.P., 1971. "Deep-Well Injection of Wastewater". *Journal of Water Pollution Control Federation*, **43(7)**: 1468-1479.

United Nation for Environment and Protection Environmental Data Report, (1991). 3rd edn., Blackwell, Oxford.

List of Contributors

1. **Agrawal, D.K. (Dr.)**

   GB Pant Institute of Himalayan Environment and Development, Kosi-Katarmal, Almora – 263 643, India
   (Presently with National Green Tribunal, Principal Bench, New Delhi)

   E-mail: devendraagrawal26@gmail.com

2. **Arora, Sanjay (Dr.)**

   Division of Soil Science and Agricultural Chemistry, S.K. University of Agricultural Sciences and Technology, FoA, Chatha – 180 009, Jammu (J&K), India

   E-mail: aroraspau@yahoo.co.in

3. **Biswas, Hrittick (Dr.)**

   ICAR-Indian Institute of Soil and Water Conservation, Research Centre, Cantonment, Bellary – 583 104, Karnataka, India

   E-mail: hritbis@yahoo.co.in

4. **de la Riva, J. (Dr.)**

   Dep. Geografía y Ordenación del Territorio, Universidad de Zaragoza, Spain

   E-mail: iriva@mncn.csic.es

5. **Dubey, Kumud (Dr.)**

Centre for Social Forestry and Eco-Rehabilitation, 337-Ashok Nagar, Allahabad – 211 001, U.P., India

E-mail: dkumud@yahoo.com

6. **Gill, A.S. (Dr.)**

Indian Grassland and Fodder Research Institute, Kanpur road, Jhansi, U.P., India

E-mail: asgill29@yahoo.com

7. **Gupta, Anil K. (Dr.)**

National Institute of Disaster Management, IIPA Campus, New Delhi – 110 002, India

E-mail: envirosafe2007@gmail.com

8. **Hadda, M.S. (Dr.)**

Department of Soils, Punjab Agricultural University, Ludhiana – 141 004, Punjab, India

9. **Hoppe, A. (Dr.)**

Institut für Angewandte Geowissenschaften, Technische Universität Darmstadt, Germany

E-mail: andreas.hoppe@charite.de

10. **Katiyar, V.S. (Dr.)**

Central Soil and Water Conservation Research and Training Institute, Datia – 475 661, M.P., India

E-mail: hritbis@yahoo.co.in

11. **Khan, Md Rizwan**

Centre for Social Forestry and Eco-Rehabilitation, 337-Ashok Nagar, Allahabad – 211 001, U.P., India

12. **Khoiyangbam, R.S. (Dr.)**

Institute of Environment and Development Studies, Bundelkhand University, Jhansi, U.P., India

E-mail: khoiyangbam@yahoo.co.in

13. **Kumar, Avinash (Dr.)**

Department of Marine Geology, Mangalore University, Mangalore, India

E-mail: kumaravinash13@rediffmail.com

14. Kumar, Vinit (Dr.)

Institute of Environment and Development Studies, Bundelkhand University, Jhansi, U.P., India

15. Lal, Brij (Dr.)

Central Soil and Water Conservation Research and Training Institute, Datia – 475 661, M.P., India

E-mail: lakaria2001@yahoo.com; hritbis@yahoo.co.in

16. Lamelas, M.T. (Dr.)

Dept. Geografía y Ordenación del Territorio, Facultad de Filosofía y Letras, Universidad de Zaragoza, Pedro Cerbuna 12, 50009, Zaragoza, Spain

E-mail: tlamelas@unizar.es, tlamelas@terra.es

17. Lodhi, M. S. (Dr.)

GB Pant Institute of Himalayan Environment and Development, Kosi-Katarmal, Almora – 263 643, India

E-mail: mahen29.mail@gmail.com

18. Lodhia, Shital (Dr.)

Department of Geography, University of Baroda, Vadodara, Gujarat, India

E-mail: s_lodhia@rediffmail.com

19. Marinoni, O. (Dr.)

CSIRO Ecosystem Sciences, Sustainable Agriculture Flagship, Dutton Park, Qld 4102, Australia

E-mail: oswald.marinoni@csiro.au

20. Mohan, S. (Prof. Dr.)

Environmental and Water Resources Engineering Division, Indian Institute of Technology, Chennai, Tamil Nadu, India

E-mail: prof_smohan@yahoo.com

21. Nair, Sreeja, S.

Disaster Risk Reduction Consortium, UN-Habitat, Yangon, Myanmar

E-mail: sreeja.unhabitat@gmail.com

22. Oinam, Santaram S.

GB Pant Institute of Himalayan Environment and Developmental Studies, Kullu, H.P., India

23. Pandey, Nisha (Dr.)

MITCON Institute of Management, Pune, Maharashtra, India

E-mail: nisha_pandey16@rediffmail.com

24. **Pandey, R.K. (Dr.)**

State Forest Research Institute, Polipathar, Jabalpur, Madhya Pradesh, India

E-mail: pandeyrk1@yahoo.com

25. **Pathak, Jharna (Dr.)**

Gujarat Institute of Development Research, Ahmedabad, Gujarat, India

E-mail: fmjharna@yahoo.com

26. **Rai (Bose), Sutapa (Dr.)**

Indian Institute of Science Education and Research, Kolkata, India

E-mail: sutapa.bose@iiserkol.ac.in

27. **Ramamohan, Y. (Dr.)**

Centre for Environment, IST, Jawaharlal Nehru Technological University, Hyderabad, A.P., India

E-mail: ramamohan_y@yahoo.com

28. **Ramani Bai, V. (Dr.)**

Department of Civil and Environmental Engg., Faculty of Engineering, University of Malaya, 50603 Kuala Lumpur, Malaysia

E-mail: ramanibai@um.edu.my

29. **Sachdev, Sanjeev (Dr.)**

Lake Conservation Authority of Madhya Pradesh, Bhopal, India

E-mail: sachdev_sanjeev@yahoo.com

30. **Saini, S.K. (Dr.)**

State Forest Research Institute, Polipathar, Jabalpur, M.P., India

31. **Singh, Gurmeet (Dr.)**

School of Environmental Sciences, Jawaharlal Nehru University, Delhi

E-mail: singh_gurmeet@hotmail.com

32. **Singh, V.K.**

Centre for Social Forestry and Eco-Rehabilitation, 337-Ashok Nagar, Allahabad – 211 001, U.P., India

33. **Srivastava, Anubha**

Centre for Social Forestry and Eco-Rehabilitation, 337-Ashok Nagar, Allahabad – 211 001, U.P., India

34. Tyagi, Pallavee (Dr.)

Department of Natural Resource Management, Institute of Environment and Development Studies, Bundelkhand University, Jhansi – 284 128, U.P., India.

E-mail: palavee1@rediffmail.com

35. Vasishth, Ashwani (Dr.)

School of Social Science and Human Services, Ramapo College, Mahwah, NJ, USA

E-mail: vasishth@rampo.edu

36. Vedak, V. (Dr.)

MITCON Institute of Management, Pune, Maharashtra, India

E-mail: directormima@gmail.com

37. Yunus, M. (Prof., Dr.)

Professor (Retd.), Ex-Head and Dean, School for Environmental Sciences, B.B. Ambedkar (Central) University, Lucknow – 226 025, U.P., India

F-45, Sarvodaya Nagar, Lucknow – 226 016, U.P., India

E-mail: profyunusm@gmail.com

38. Yusoff, Sumiani (Dr.)

Department of Civil and Environmental Engineering, Faculty of Engineering, University of Malaya, 50603 Kuala Lumpur, Malaysia

E-mail: sumiani@um.edu.my

Index

M

N

O

P

R

S

T

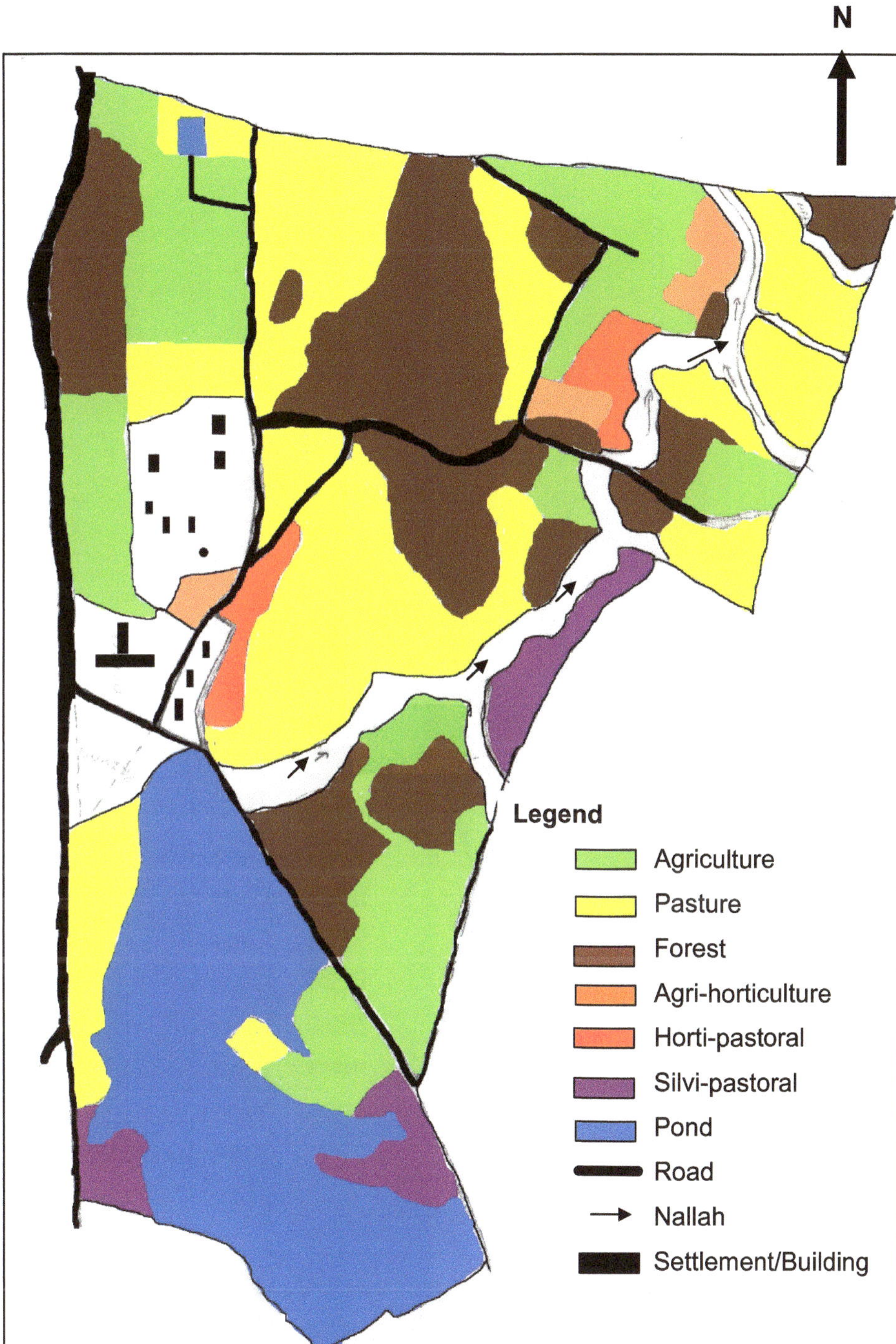

Figure 2.1: Present land-use map of the study area (191 ha) **(P.17)**

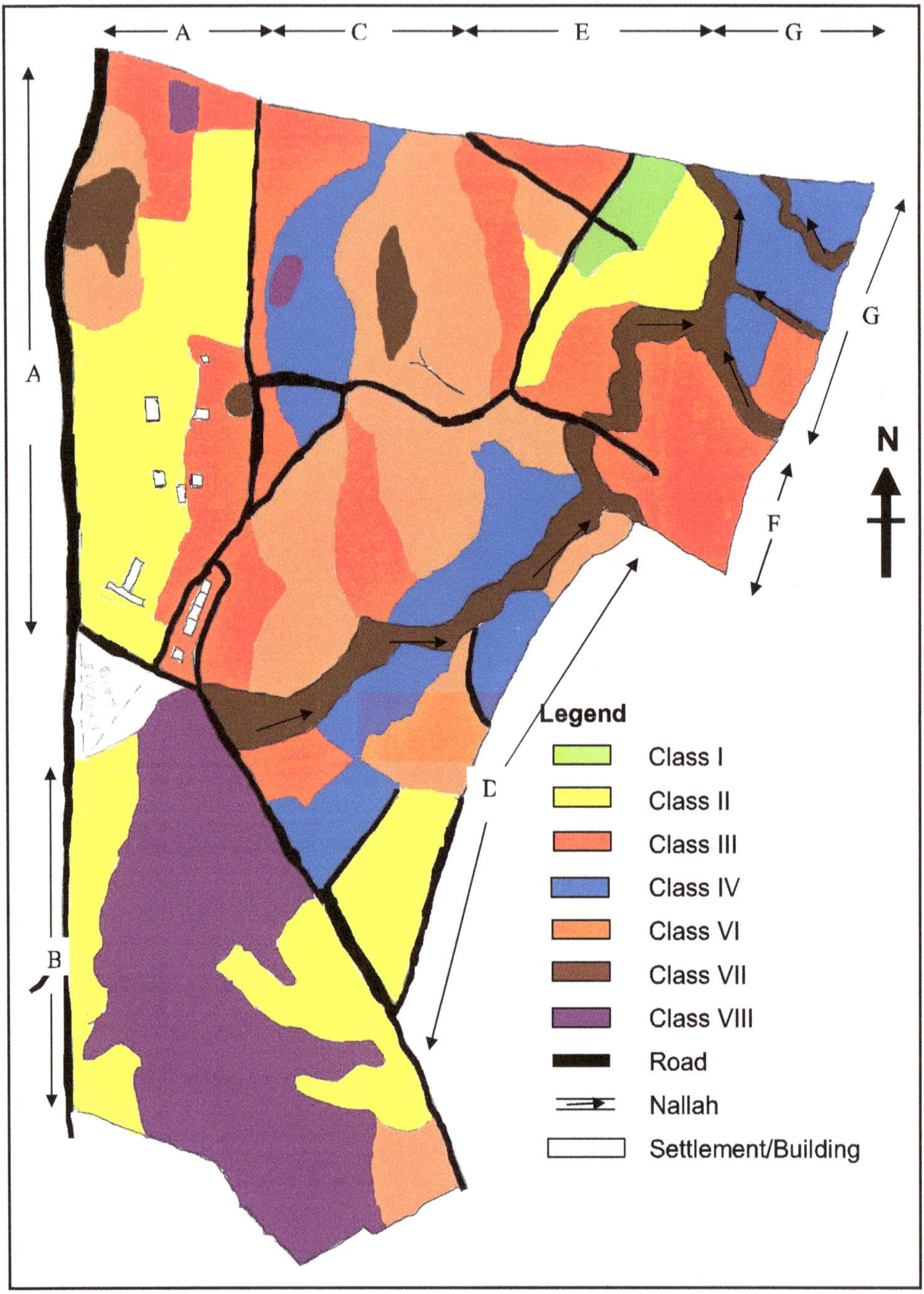

Figure 2.2: Land capability classification map of CSWCRTI, Research Farm, Datia (191 ha) **(P.18)**

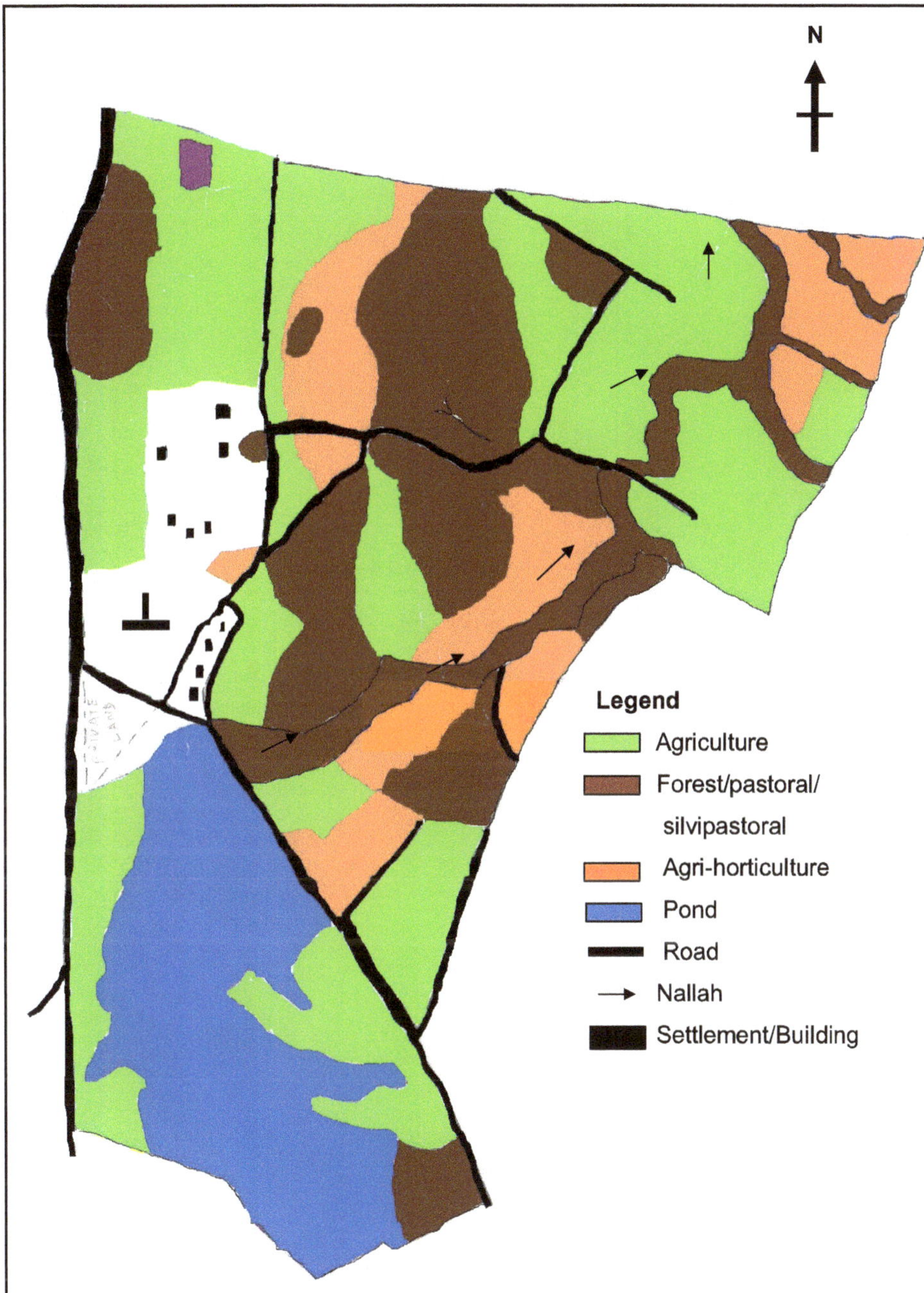

Figure 2.3: Proposed land-use of the farm **(P.19)**

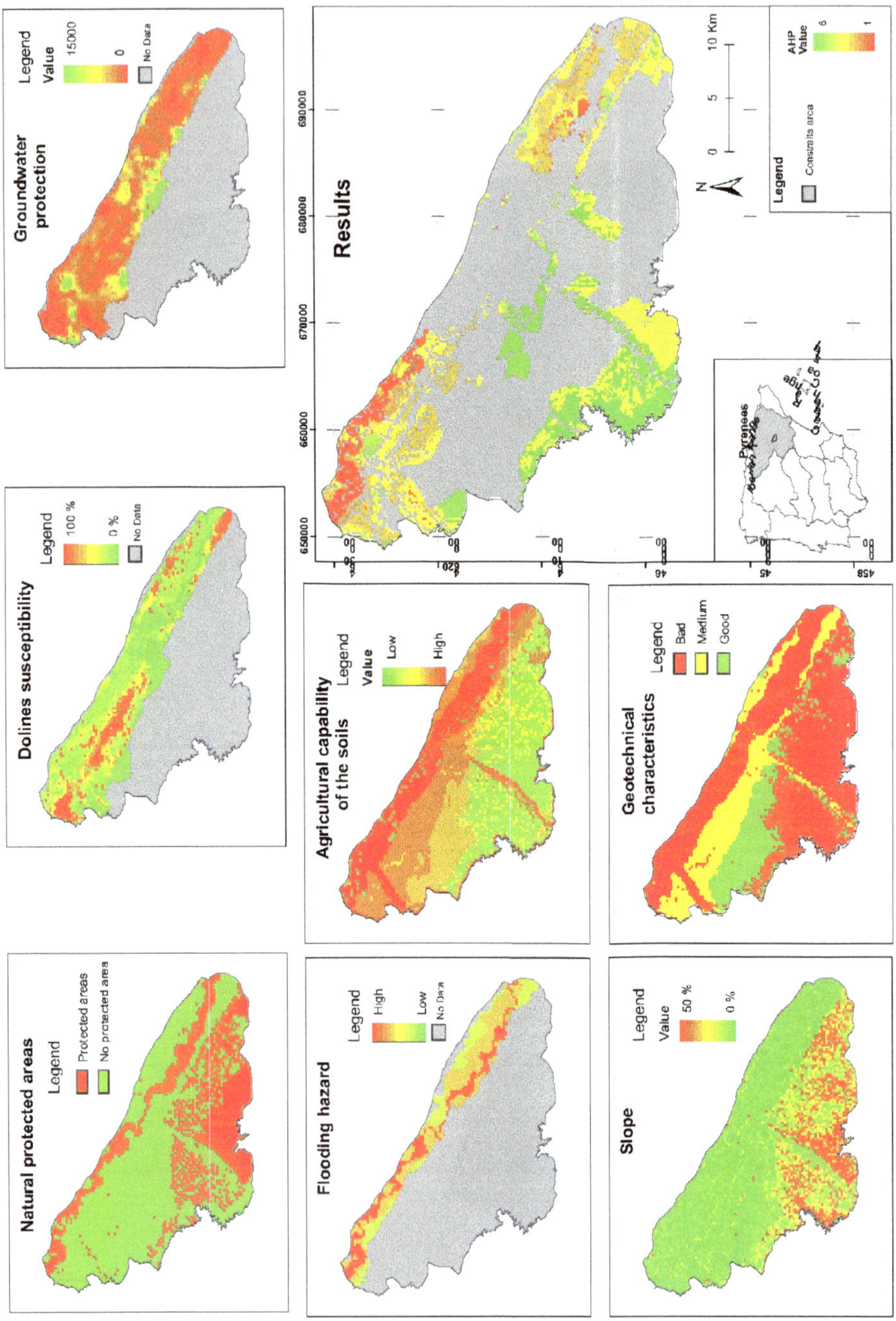

Figure 5.2: Results from the industrial use suitability analysis and original values**(P.59)**

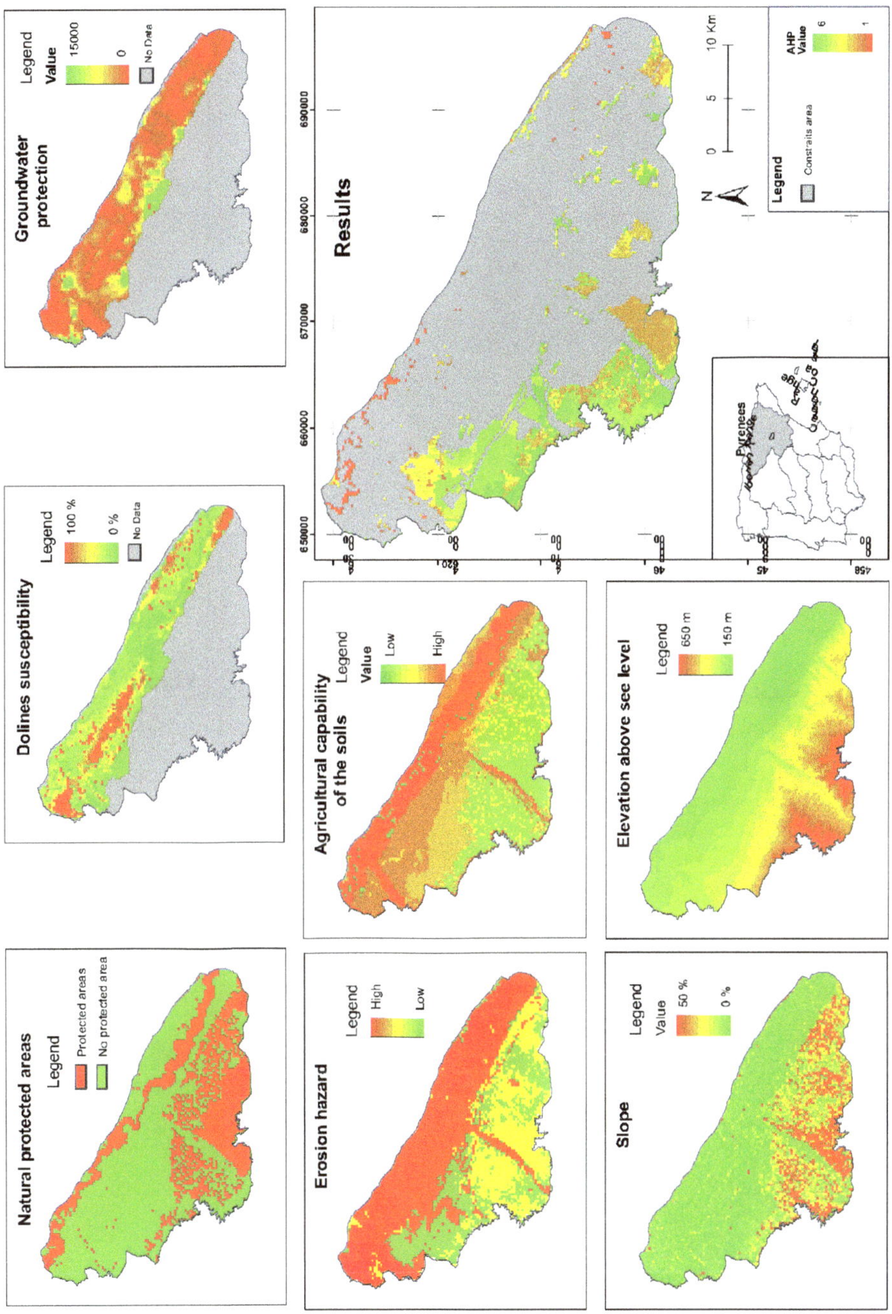

Figure 5.4: Irrigated agricultural use suitability**(P.63)**

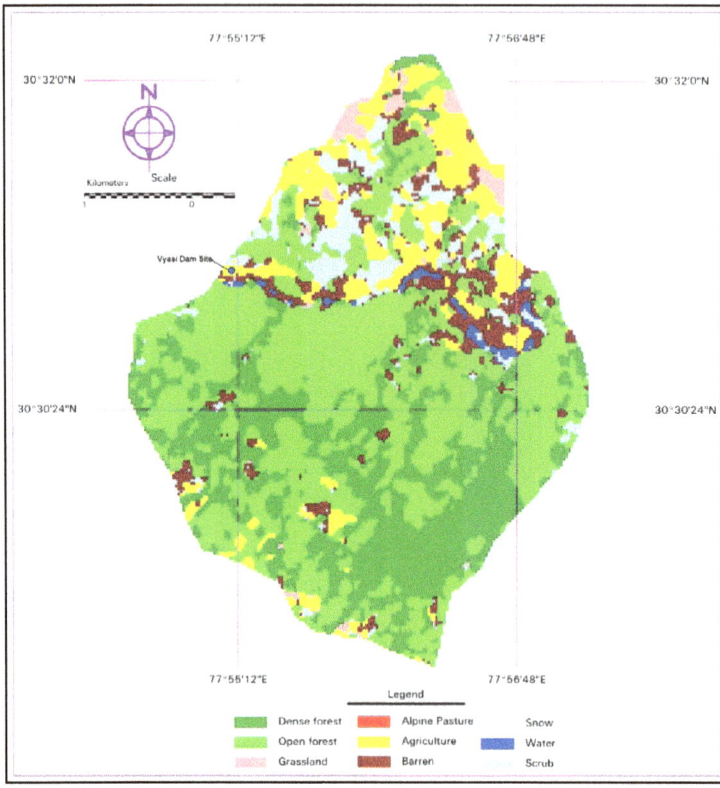

Figure 8.2: **(P.97)**

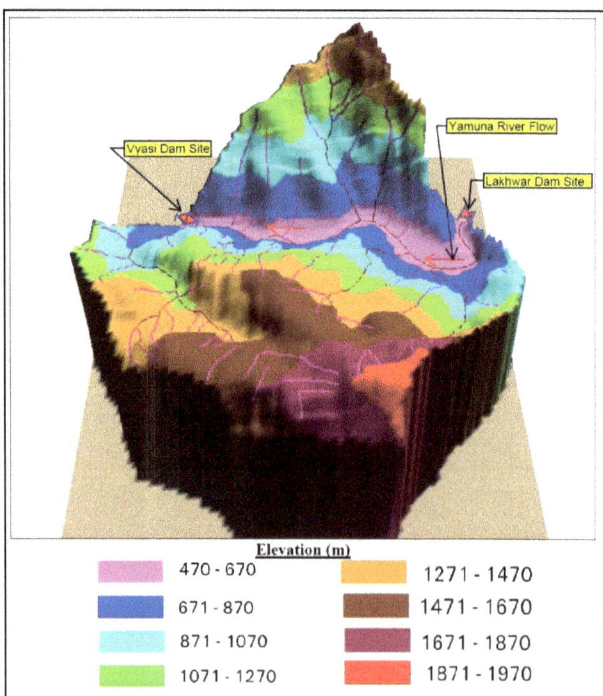

Figure 8.3: **(P.99)**

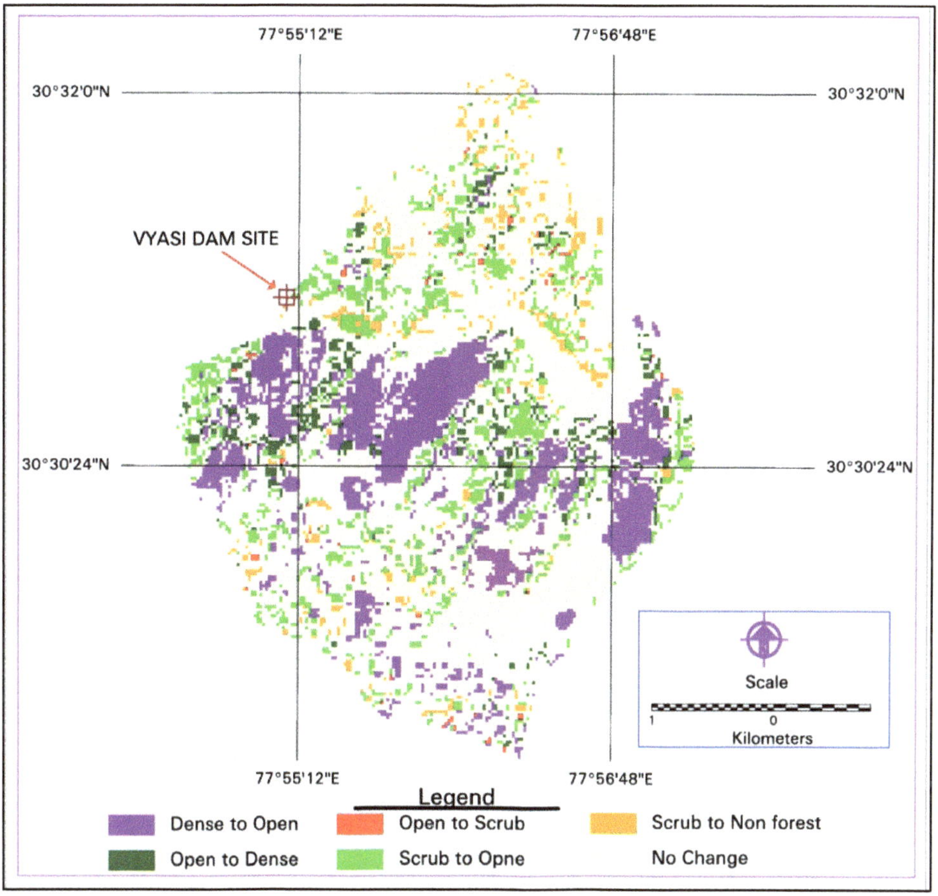

Figure 8.5: Land-use/land-cover changes over a period of 12 years (1991-2002) in the free draining catchment of Vyasi HEP **(P.101)**

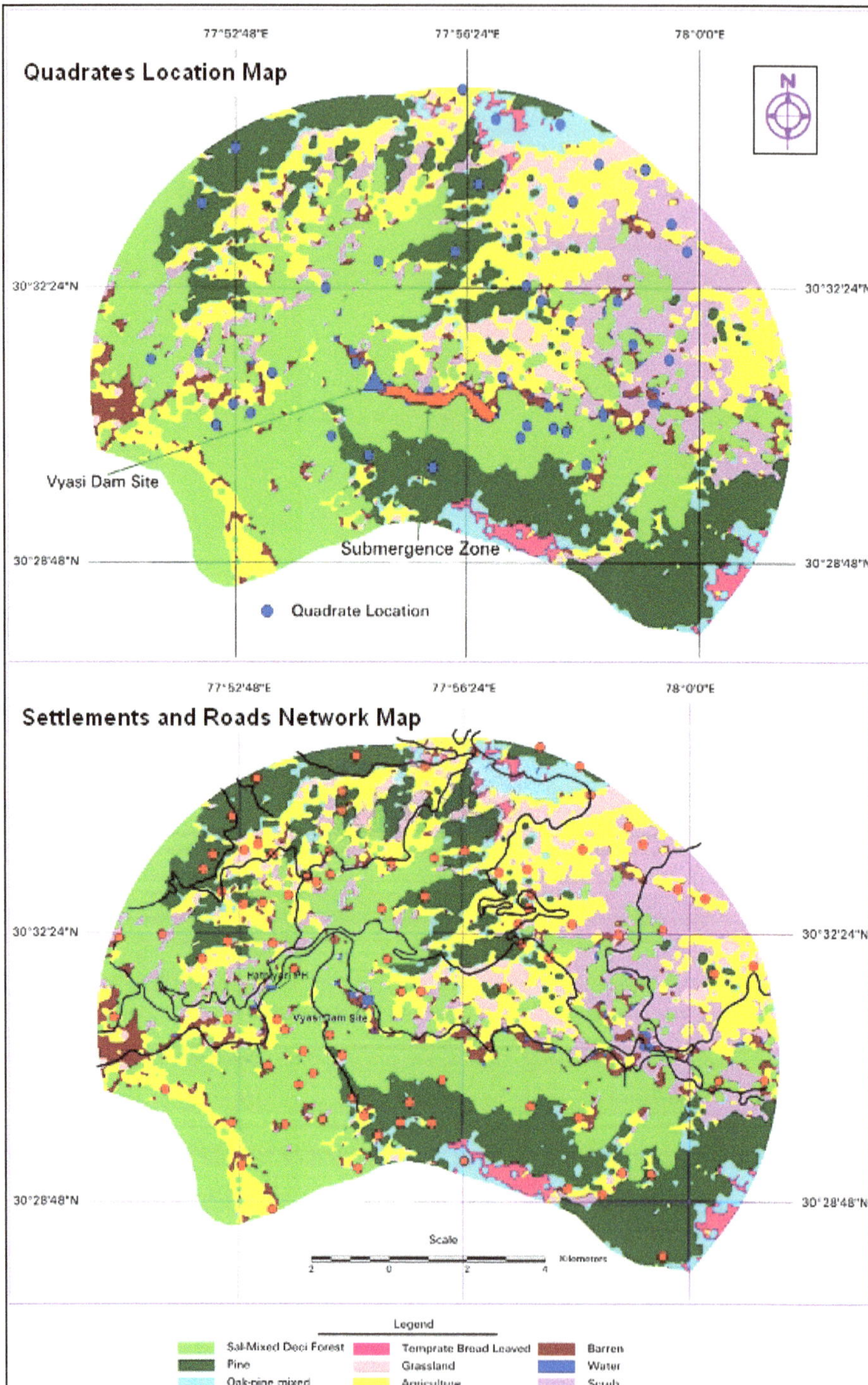

Figure 8.6: Land-use/land-cover map and location map of quadrates used for biodiversity assessment of the influence zone of Vyasi HEP along with roads network and settlements **(P.102)**

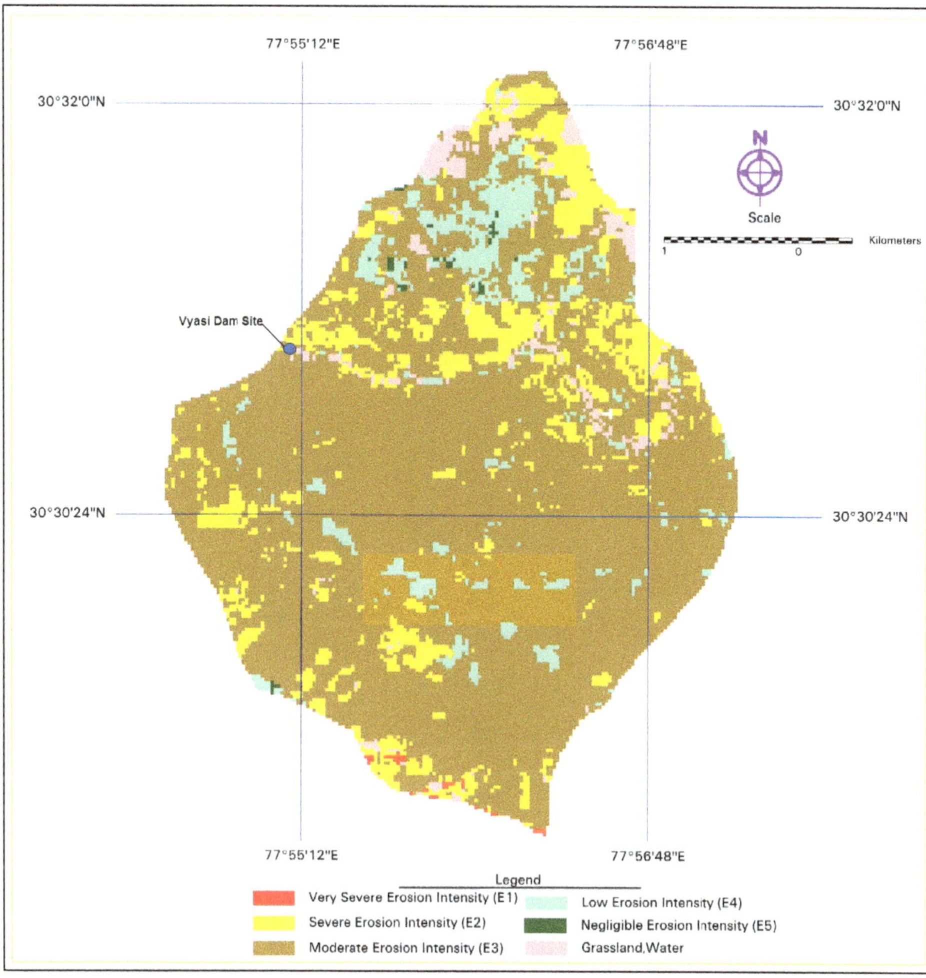

Figure 8.9: Composite erosion intensity map of the free draining catchment of Vyasi HEP **(P.107)**

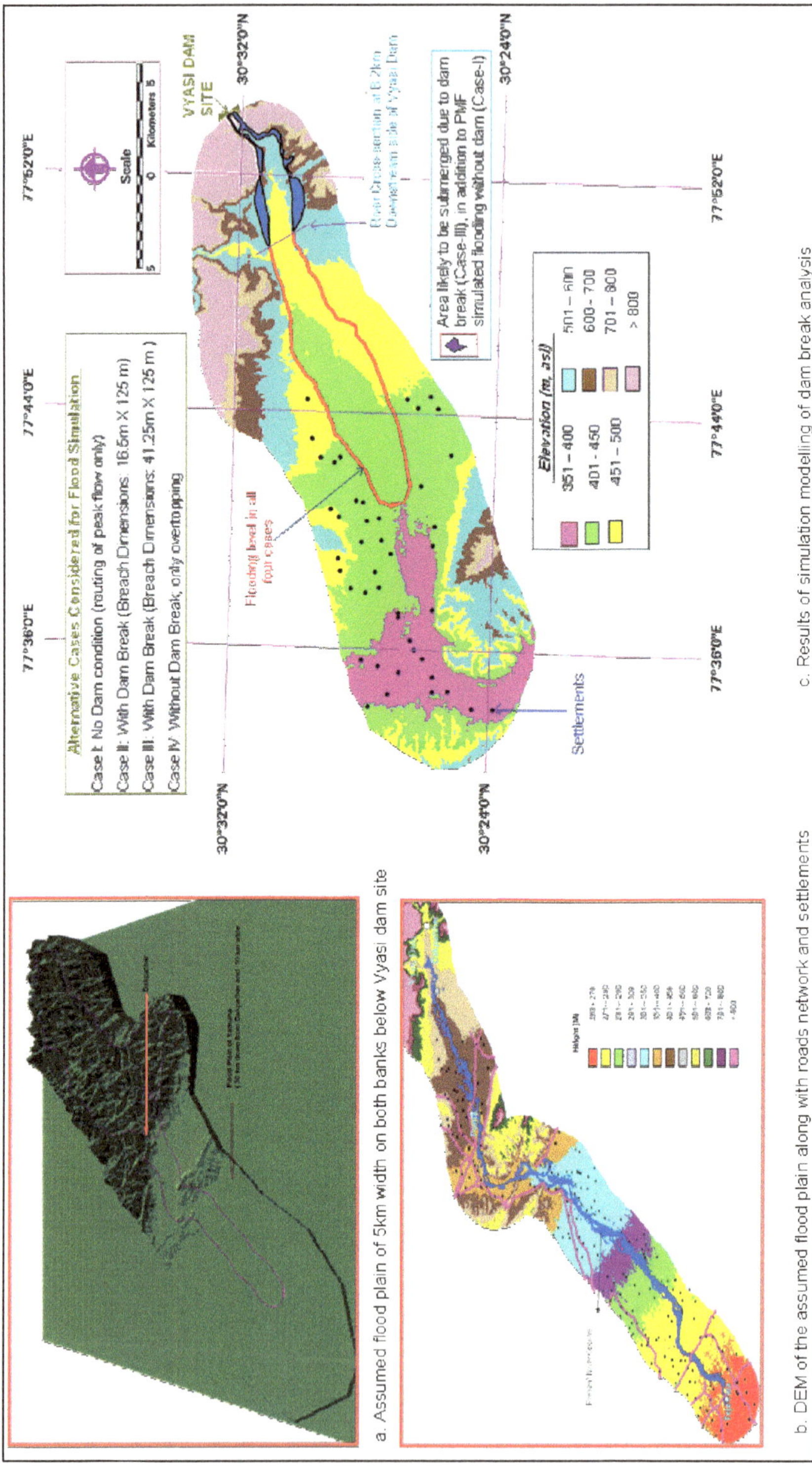

a. Assumed flood plain of 5km width on both banks below Vyasi dam site

b. DEM of the assumed flood plain along with roads network and settlements

c. Results of simulation modelling of dam break analysis

Figure 8.11: Flood inundation map in assumed floodplain of 10km width in downstream of Vyasi HEP **(P.109)**

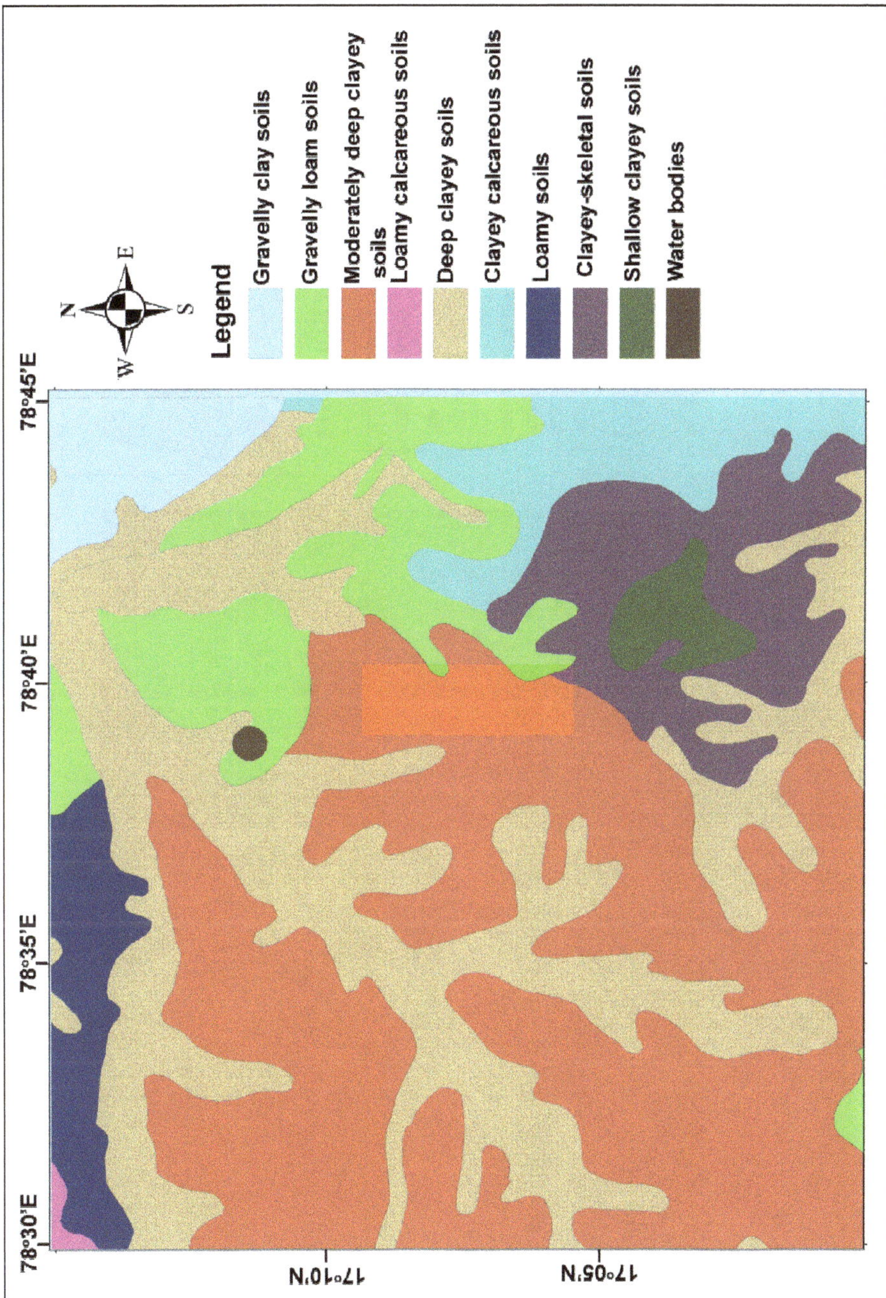

Figure 9.2: Soil map of the study area showing the distribution of various types **(P.116)**

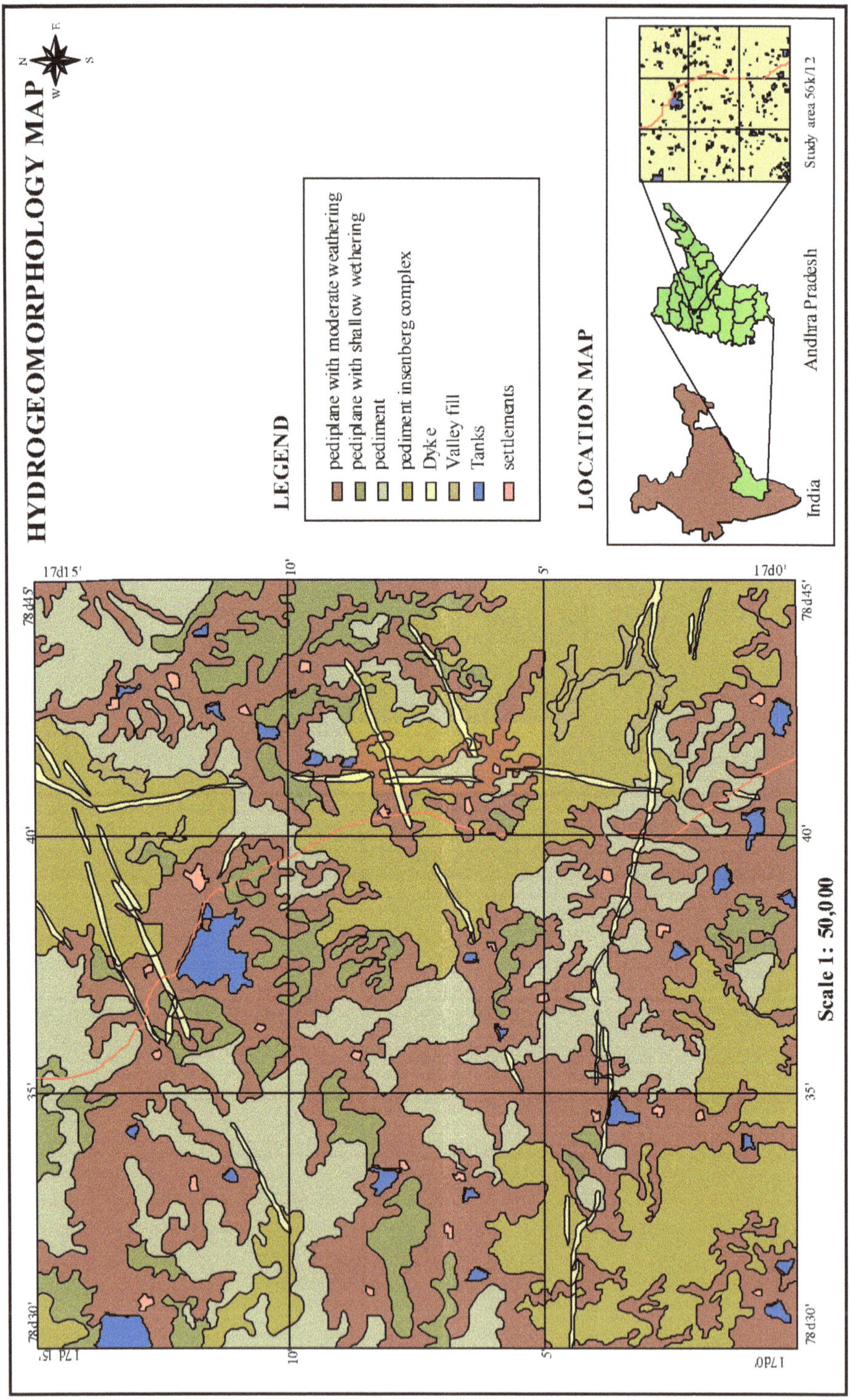

Figure 9.4: The hydrogeomorphic map showing various geomorphologic landforms **(P.119)**

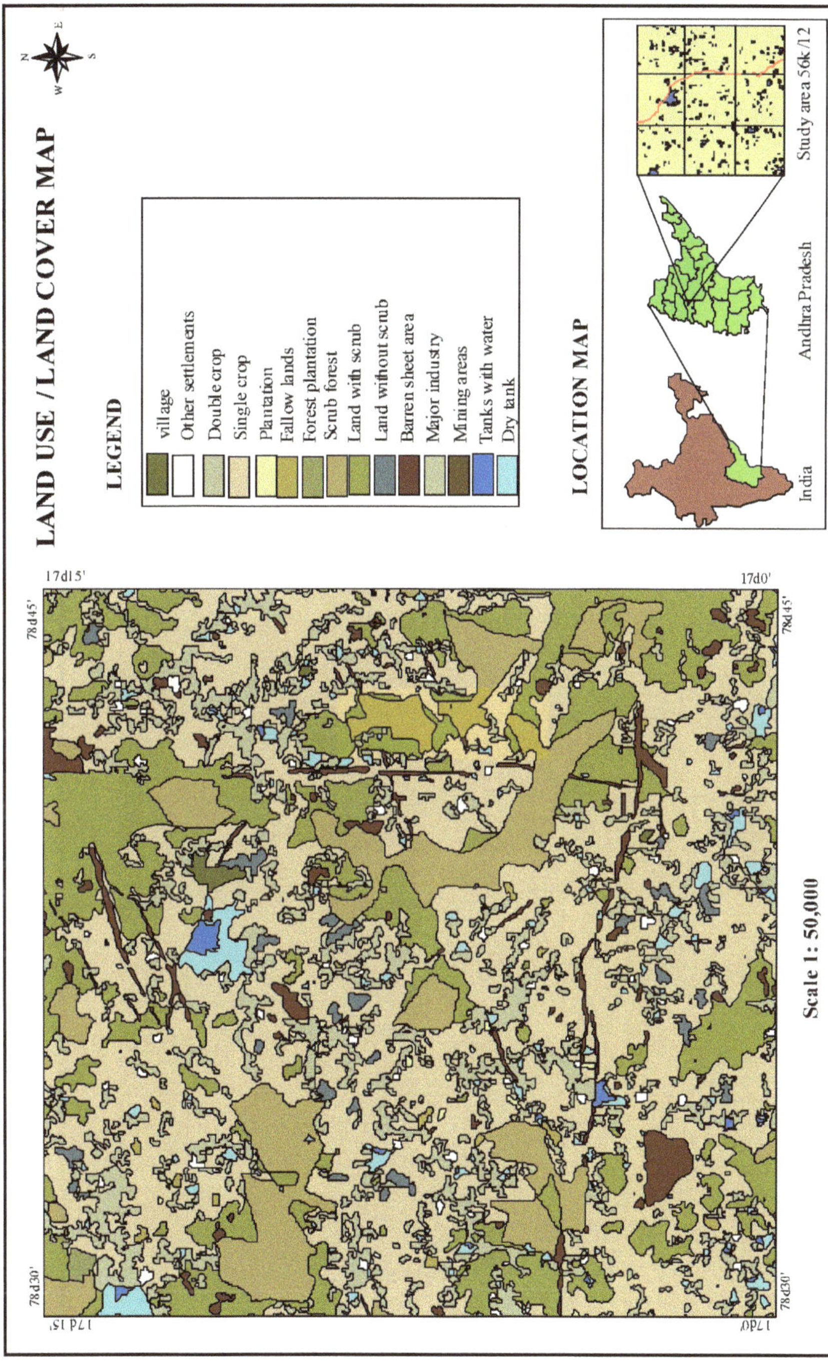

Figure 9.5: The map showing the land use/land cover distribution of the study area **(P.120)**

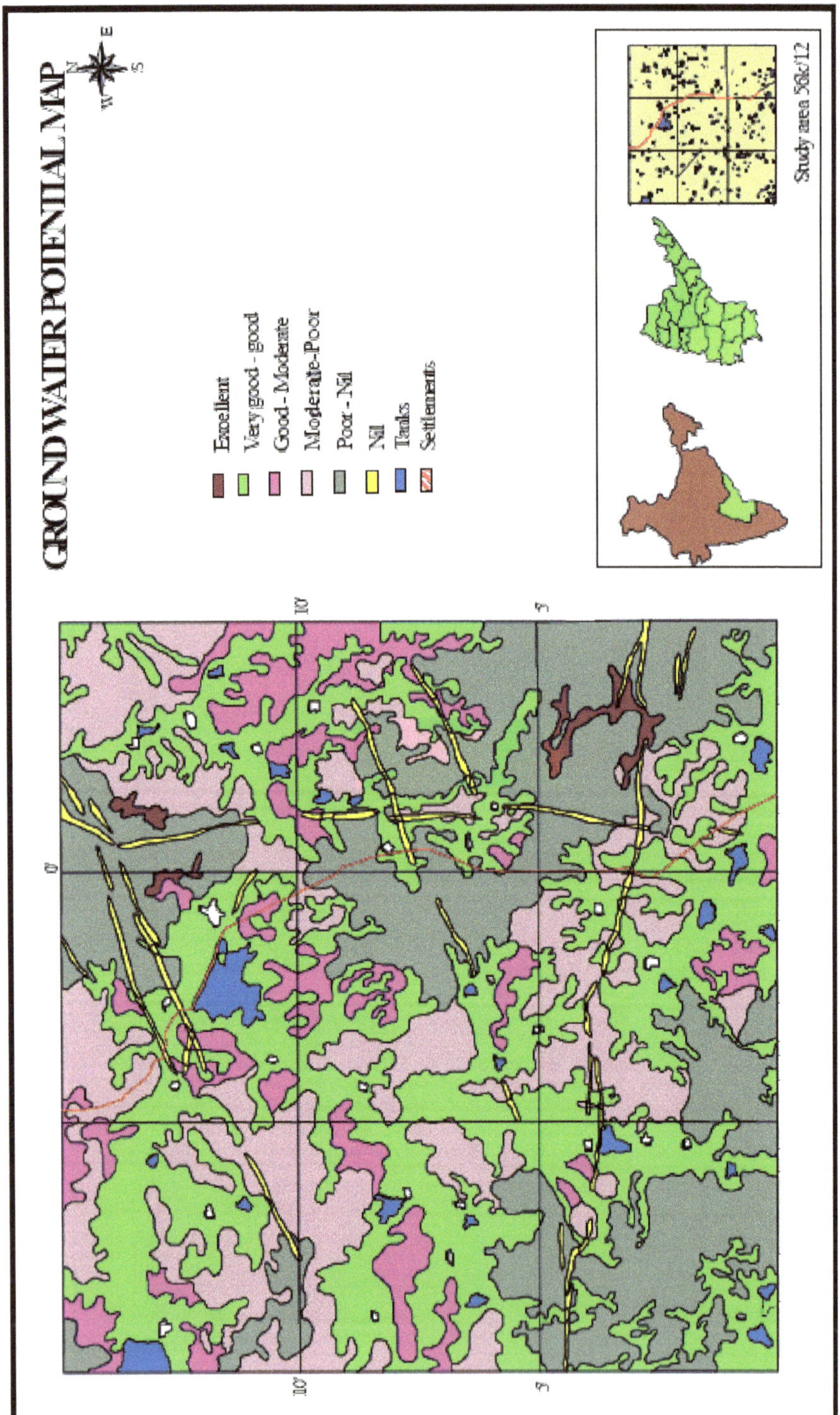

Figure 9.6: Groundwater potential map of the study area **(P.121)**

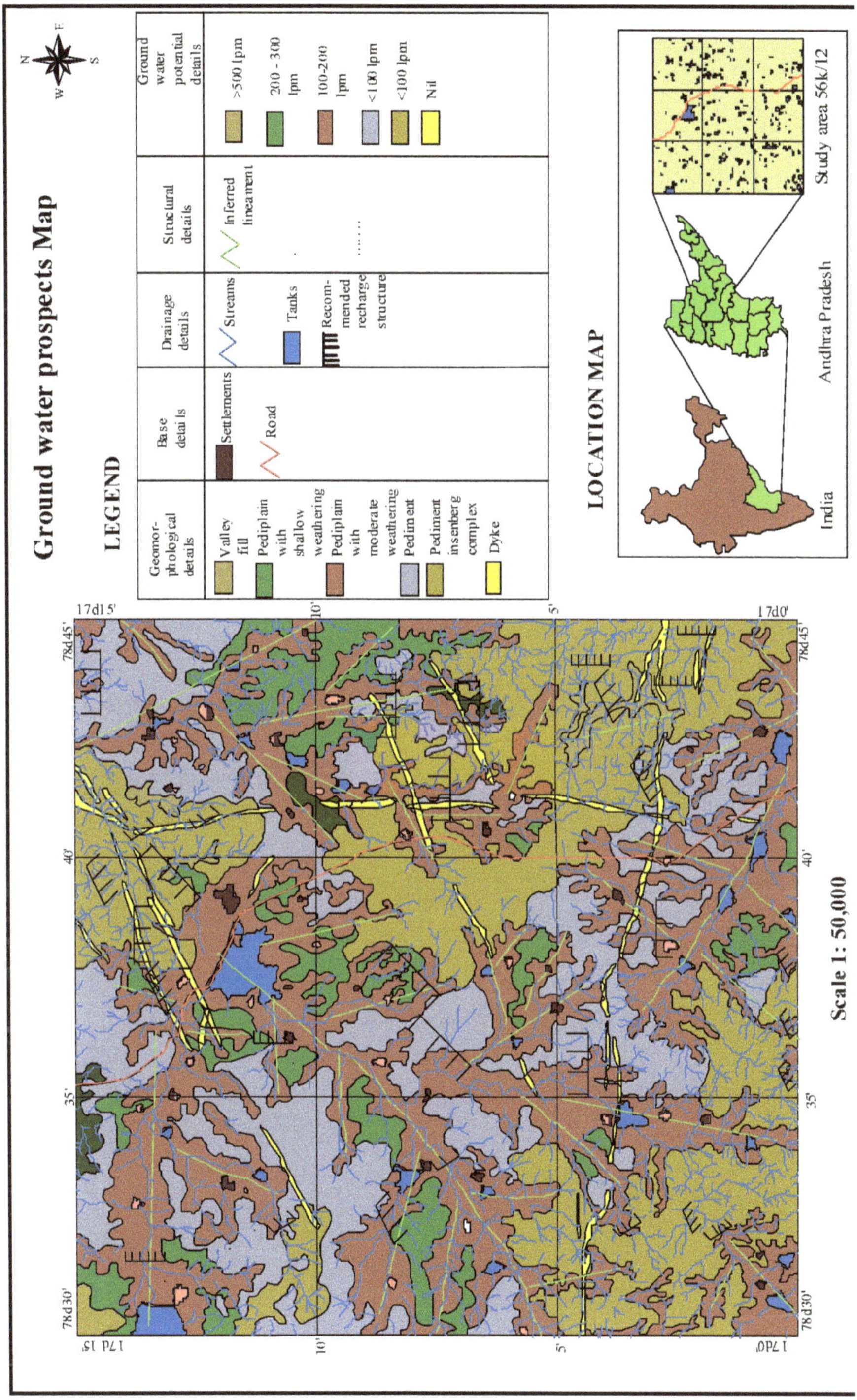

Figure 9.7:Groundwater prospects map of the study area **(P.122)**

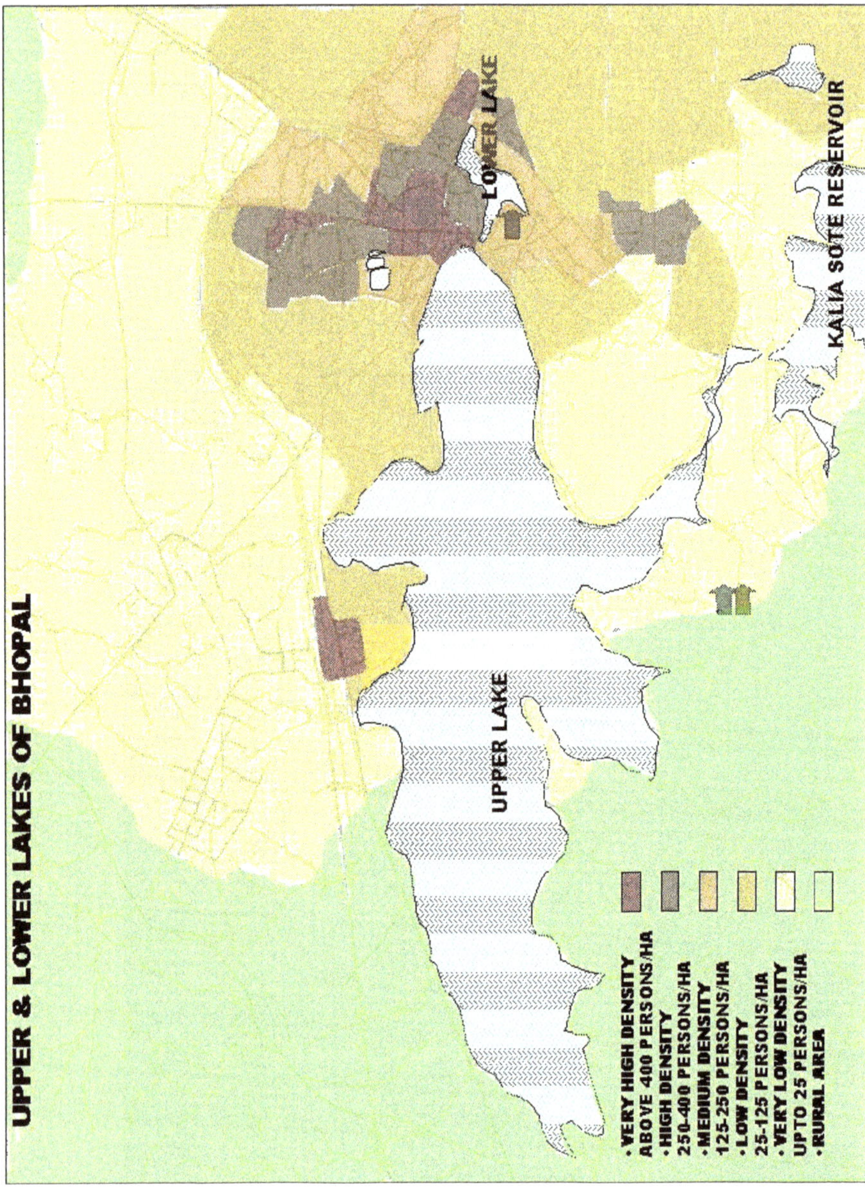

Figure 16.1: Upper and lower lakes of Bhopal **(P.200)**

www.ingramcontent.com/pod-product-compliance
Ingram Content Group UK Ltd.
Pitfield, Milton Keynes, MK11 3LW, UK
UKHW022212051026
15624UKWH00001B/129

9 789351 240891